Resistance in the Deceleration Lane

LITERARY AND CULTURAL THEORY

General Editor: Wojciech H. Kalaga

VOLUME 42

Marzena Kubisz

Resistance in the Deceleration Lane

Velocentrism, Slow Culture and Everyday Practice

Bibliographic Information published by the Deutsche Nationalbibliothek
The Deutsche Nationalbibliothek lists this publication in the Deutsche Nationalbibliografie; detailed bibliographic data is available in the internet at http://dnb.d-nb.de.

Library of Congress Cataloging-in-Publication Data
Kubisz, Marzena.
  Resistance in the deceleration lane : velocentrism, slow culture and everyday practice / Marzena Kubisz. – Peter Lang Edition.
     pages cm. – (Literary and cultural theory ; volume 42)
  Includes bibliographical references.
  ISBN 978-3-631-65558-0 (Print) – ISBN 978-3-653-04739-4 (E-Book)
  1. Time–Sociological aspects. 2. Speed–Social aspects. 3. Slow life movement. 4. Technological innovations–Social aspects. I. Title.
  HM656.K83 2014
  304.2'3–dc23
                                                           2014029542

This Publication was financially supported
by the University of Silesia.

ISSN 1434-0313
ISBN 978-3-631-65558-0 (Print)
E-ISBN 978-3-653-04739-4 (E-Book)
DOI 10.3726/978-3-653-04739-4

© Peter Lang GmbH
Internationaler Verlag der Wissenschaften
Frankfurt am Main 2014
All rights reserved.
Peter Lang Edition is an Imprint of Peter Lang GmbH.

Peter Lang – Frankfurt am Main · Bern · Bruxelles · New York ·
Oxford · Warszawa · Wien

All parts of this publication are protected by copyright. Any utilisation outside the strict limits of the copyright law, without the permission of the publisher, is forbidden and liable to prosecution. This applies in particular to reproductions, translations, microfilming, and storage and processing in electronic retrieval systems.

This publication has been peer reviewed.

www.peterlang.com

For J.K.

Only connect…
E. M. Forster, *Howards End*

Contents

Introduction: Between Mercury and Jove 11

Part One
Speed: Towards Velocentric Perspective 21

Chapter One
"To Fly like Mercury": Acceleration and Modern Experience 23
Chapter Two
Streamlined Culture and the Rise of the Mis-man 47

Part Two
Slow Time and Alternative Hedonism 73

Chapter Three
"To Sit like Jove": Slowness and Everyday Resistance 75
Chapter Four
The Temporality of (Other) Pleasures 99

Part Three
The Other Speed: Cultural Practices and Representations 123

Chapter Five
New Territoriality in the Age of Deterritorialization 125
Chapter Six
Negotiating Mobility: On the Slow Move 149
Chapter Seven
"Relocation, Relocation": Slow Goes Pop 175

Conclusions: Post-Slow? Between Strategies and Props 203

Bibliography 209

Introduction: Between Mercury and Jove

> That [deceleration lane] allows you to decelerate to the proper speed, so as not to exit the freeway or interstate at an unsafe speed, which could result in a possible rollover of your vehicle.
> Reed Berry, *Highway Driving Safety*

In a letter to John Hamilton Reynolds, John Keats wrote:

> Now it is more noble to sit like Jove than to fly like Mercury – let us not therefore go hurrying about and collecting honey-bee like, buzzing here and there impatiently from a knowledge of what is to be arrived at; but let us open our leaves like a flower and be passive and receptive, budding patiently under the eye of Apollo and taking hints from every noble insect that favours us with a visit. Sap will be given us for Meat and dew for drink. I was led into these thoughts, my dear Reynolds, by the beauty of the morning operating on a sense of Idleness. I have not read any Books. The Morning said I was right. I had no Idea but of the Morning and the Thrush said I was right [...].[1]

The year 1818, when Keats wrote this letter to his friend, was perhaps one of the last moments when the poet's words of encouragement to be idle and passive might still have resonated with remains of the appreciation which Western culture had developed for idleness. The rise of the middle classes and their work ethos, coupled with the acceleration of everyday life brought on by the Industrial Revolution and technological progress, limited social acceptance for idleness and equated efficiency with quickness. Read in the context of the history of the culture of speed the above excerpt illustrates the co-existence of two speed-oriented perspectives: the Mercurial perspective of "hurrying" and "buzzing" is counterbalanced by the Jovian perspective which turns being "passive and receptive" into a desirable state, highly dependent on one's ability to dwell in contemplation. Keats' appreciation of the moment of retreat from a world of hurry was not intended to undermine or discredit active participation in its dynamic flows and movements but rather to emphasize a need to slow down in order to balance and regulate the rhythm of life, which by many people was seen as increasingly unbalanced and deregulated. This book, in its widest grasp, is about the relation between these two perspectives.

While it is true to say that the Mercurial principle has dominated modern experience and monopolized time management in affluent societies, the Jovian

---

1 John Keats, "To J. K. Reynolds," in *Selected Letters of John Keats*, ed. Grant F. Scott (Cambridge, Mass.: Harvard University Press, 2005), p. 93.

aspect of the culture of speed has recently started to mark its presence in various spheres of life and generate a wide array of cultural texts and practices which are informed by a desire to rehabilitate slowness. A growing realization that speed has become "a bit of a troublesome servant that comes more often than it is called," to quote Valéry Larbaud (who in 1930 in a manner most prophetic predicted that one day speed would become man's "servant-mistress"[2]), has resulted in an explosion of interest in the consequences of the development of technologies of speed and in the chance to confront them by reinventing one's relation to time. Although Wendy Parkins is right to say that "living slowly as a form of rejecting the cultural orthodoxy of speed is not a recent phenomenon,"[3] the popularity of the narratives of slow culture has never reached such a scale before; slow practices which are popular today have never been marked by such dynamics nor have they struck such a chord in so many different areas. Slow living has entered the consciousness of contemporary man as a postulate, a utopian dream and a more or less formalized practice.

The forms that the responses to fast life and its requirements take are diversified and although they are all embraced by one term, i.e. "slow living," it has to be stressed that they are far from being homogenous in terms of motives, aspirations, practices and interests. The concept of cultural slowness is used throughout this book to theorize not only those practices and movements that are immediately associated with the idea of slowing down, such as the Slow Food or Slow Cities movements, but also to include practices which, although not performed under the banner of a "slow revolution," valorize deceleration as a powerful cultural principle and acknowledge its potential to change the quality of both the individual and collective experience of living in an accelerated culture. The first case in point here is the so-called "copenhagization," a large-scale urban project inspired by a desire to reclaim the space of the city that has been taken over by car culture. The second one is downshifting, which is underpinned by the belief that "less is more," to use what might be seen as the motto of all practices which challenge a popular belief that happiness and satisfaction in consumer culture are determined by consumers' spending powers. All the texts and practices of slow culture analyzed in this book under the rubric of the "slow" postulate, although in different ways and to varying degrees, the possibility of negotiating the speed contract, the signing of which modern culture has

---

2   Valéry Larbaud, "Slowness," in *Speed Limits*, ed. Jeffrey T. Schnapp, trans. Christy Wampole (Milan: Skira, 2009), p. 276.
3   Wendy Parkins, "Out of Time: Fast Subjects and Slow Living," *Time & Society*, Vol. 13 No. 2/3 (2004), p. 365.

construed as indispensable for success, efficiency and the good life. They are seen as narratives which aim at creating a certain vision of the world, provide their adherents with sets of meanings and points of reference and help them achieve "[p]ersonal integrity, as the achievement of an authentic self," which, as Anthony Giddens claims, "comes from integrating life experiences within the narrative of self-development."[4]

Symptomatic of the growing realization of the cultural significance of deceleration in recent years is the publication of numerous works which provide a consistent mode of thinking through the phenomenon of slow living and help identify the areas and cultural practices affected by the discourse of slow culture. Two works deserve special mention for the role they have played in the systematization of theoretical reflections and critical perspectives concerning slowness: Wendy Parkins and Geoffrey Craig's *Slow Living* (2006), which focuses primarily on organized social movements and their resistive potential, and a collection of essays edited by Nick Osbaldiston, entitled *Culture of the Slow. Social Deceleration in an Accelerated World* (2013), which takes the analysis of slow life far beyond organized movements and the moral and ethical motivations of their members. Drawing on the scholarship and research of Parkins, Craig, Osbaldiston and other cultural analysts and theorists of deceleration, the present study places critical emphasis on different aspects of the phenomenon in question. The version of slow culture which this book attempts to construct is placed within the parameters of cultural studies and their interest in the ways in which values, social norms and conventions are produces and represented in particular societies and communities, and in this respect it shares a critical perspective with Parkins, Craig and Osbaldiston. What makes it different, though, is that the book addresses slow living at two levels: that of macro and of micro analysis. In its macro-dimension it seeks to examine slowness in a broader context of the culture of speed and see them, following Jeffrey T. Schnapp's postulate, not as "polar opposites," but as "dual emanations of a single system within which speed is king."[5] My interest lies first in how particular meanings got constructed around the concept of speed and achieved "the status of 'commonsense'" and "a certain taken-for-granted quality"[6] and, second, in placing the phenomena generated by

---

4 Anthony Giddens, *Modernity and Self-Identity. Self and Society in the Late Modern Age* (Stanford, CA.: Stanford University Press, 1991), p. 80.
5 Jeffrey T. Schnapp, "Fast (Slow) Modern," in *Speed Limits*, ed. Jeffrey T. Schnapp (Milan: Skira, 2009), p. 27.
6 John Storey, "Introduction: The Study of Popular Culture and Cultural Studies," in *Cultural Theory and Popular Culture: A Reader*, ed. John Storey (Harlow: Pearson Education Limited, 1994), p. xviii.

a reactivated Jovian principle in the context of the historical evolution of the relation between man and speed. In its micro-analytical dimension the book explores the resistive and culture-forming potential of slowness on the basis of the analysis of particular texts and practices as well as their representations and the rhetoric they use. This allows me to approach the issues of resistance not only through the prism of the bipolarity of the mainstream (the dominating culture) and its outside (the spaces of resistance) but also in the context of negotiation and exchange between them. Another aspect of the analysis of slow living which the book undertakes is the rise of the mythology of slow life, which often aspires to create a holistic grand narrative which may provide an individual with stable ontological frameworks within which he can successfully reconstruct the trajectories of his own biography.

The phenomenon of slow living is not homogeneous and does not develop in the same ways in different geographic and cultural contexts. Japanese and Polish versions of slow life may stem from similar concerns, British and Hungarian models of slowing down may be motivated by similar goals, but due to different historical and cultural traditions and differing levels of affluence, slow movements adopt varied strategies and emphasize various aspects of deceleration. Although I believe that there is an interesting analytical potential in the study and comparison of various geographically inflected models of cultural deceleration, which might become a subject of further research, this book does not attempt to undertake such an analysis. Rather, its focus is on deceleration analyzed in the contexts of its origins, postulates and representations, which, regardless of their "regional" differences, permit one to see slowness as part of the history of the culture of acceleration and approach it from the perspective of continuity and change in man's relation with speed.[7]

The book is largely concerned with comprehensive analysis of practices and representations of deceleration in everyday life because this is where slow living philosophy manifests itself most visibly. However, as Ben Highmore rightly observes, any consideration of everyday life should be informed by "an immediate

---

7   This study does not examine the relation between gender and the sense of time – for stylistic reasons I refer to the "slow *man*" – but an analysis of differences in the ways in which the experience of the fast and slow time is shaped by cultural constructions of gender may offer an interesting extension of the analytical scope of slow living research. See, for example, Stefan Klein, who says: "[u]nder seemingly identical external circumstances, women evidently feel a greater sense of time pressure because they are women." Stefan Klein, *The Secret Pulse of Time. Making Sense of Life's Scarcest Commodity*, trans. Shelley Frish (Frankfurt am Main: Da Capo Press, 2007), p. 207.

question: whose everyday life?"[8] The focus here is thus on everyday cultural practices of a social group whose class identification and financial status are clearly identifiable. The phenomenon of slow living is the product of affluent Euro-American middle classes and their response to particular aspects of life, and as such needs to be examined with reference to class-specific orientations, their economic contexts and the access which the members of these classes have to technologies of consumption, production and communication. The fact that slow life is a class product, which reflects class anxieties and concerns and is marked by class-generated tensions and contradictions, should not obscure the fact that it is still a manifestation of a certain cultural condition shaped by the character of the middle-class confrontation with and participation in the culture of immediacy.

That a book about slowness should start with speed suggests the relational character of slowness and speed. Wendy Parkins claims that "the experience and value of slowness was historically derived from, and articulated through, notions of speed [...]; speed created slowness."[9] This mode of reconceptualization of fast and slow time and their mutual relation marks the methodological framework within which this study is placed. One of my broadest claims in this book is that any discussion of the rise and evolution of the *slow* man and the impact of deceleration on the present cultural moment needs to be embedded in an analysis of the ontological and epistemological conditions of the *accelerated* man and his understanding of the role speed plays in the construction of the everyday experience. Were it not for his rapid "maturation" and the dynamic expansion of the lifestyle which this process promoted, the slow man would never have become the important vehicle of cultural resistance he is now. It is only in relation to the culture of the fast that slowness displays its resistive and culture-forming potential. Psychologists Philip Zimbardo and John Boyd identify three elements that shape and determine one's temporal experience, and they claim that "[y]our emotional state, personal time perspective, and the pace of life of the community in which you live all influence the way in which you experience time."[10] It is not too much to say that in order to understand alternative temporal realities created in order to escape from the pressures and limitations of dominating models of time experience, to understand their inner tensions, contradictions

---

8   Ben Highmore, "Introduction: Questioning Everyday Life," in *The Everyday Life Reader*, ed. Ben Highmore (London: Routledge, 2002), p. 1.
9   Parkins, "Out of Time: Fast Subjects and Slow Living," p. 365.
10 Philip Zimbardo, John Boyd, *The Time Paradox. The New Psychology of Time* (London: Rider, 2008), p. 12.

and aspirations and to be able to locate them in a broad cultural panorama of contemporary society, the focus on the origins of "the pace of life of the community" seems to be an essential starting point. "Unless we understand how speed functions, what it adds and what it removes, we are deprived of the opportunity to retain slowness where it is necessary,"[11] says Thomas Hylland Eriksen.

For methodological convenience, the book is divided into three parts. Part One consists of two chapters in which I bring together elements of the history of technological acceleration. In doing this I have noted the possibility of laying down the foundations for a distinction of the historical eras which I find critical for the construction of a theoretical frame within which I place the analysis of slow life. Filtered through the prism of changing attitudes towards speed, the history of acceleration presented in the first and second chapters is subject to periodization, which helps identify the stages in the history of conceptualizations of speed and its representations. An analysis of the development of transport and communication technologies and changes in the relation between man and speed through the ages leads to a classification of three categories of accelerated man and three epochs whose characters are marked by the role speed and acceleration play in the formations of social, cultural and political relations. Thus Chapter One concentrates on the age of the running start (17th and 18th centuries) and the age of acceleration (19th century), which became, respectively, the stages of the birth of the "fast" and "fast-forward" man. The chapter illustrates the ways in which the advance of new technologies of mobility and communication influenced the modern understanding of speed and gradual domestication and naturalization of speed in the experience of everyday life. In Chapter Two I look at the 20th century, which, as John Tomlinson claims, is marked by the principle of instantaneity and which I propose to approach as an example of streamlined culture. While the "fast" man was the product of the age of coaches and the "fast-forward" man that of the age of trains and cars, the *mis-man* – multifunctional, instantaneous and simultaneous man – is represented in this study as a human type which has monopolized everyday life at the turn of the 21st century. Its distinctiveness and dominance in the spheres of social, cultural and political life of high modernity, which Anthony Giddens sees as marked by "widespread scepticism about providential reason, coupled with the recognition that science and technology are double-edged,"[12] has set the norms for cultural

---

11 Thomas Hyllard Eriksen, *Tyranny of the Moment. Fast and Slow Time in the Information Age* (London: Pluto Press, 2001), p. 59.
12 Anthony Giddens, *Modernity and Self Identity. Self and Society in the Late Modern Age* (Stanford, CA.: Stanford University Press, 1991), pp. 27–28.

practices and may be approached as a point of reference for further analysis of man's relation with speed. The analysis carried out in the first two chapters allows me to formulate the main assumptions of the velocentric perspective which I use throughout the book to reflect critically on the ways in which speeds, both high and low, and their representations affect the construction, deconstruction and reconstruction of meanings around particular cultural texts, images and practices and to register the centrality of speed in modern culture while accentuating its formative character.

The second part of the book, entitled "Slow Time and Alternative Hedonism," brings a change of scale to the analysis and a shift of the analytical spotlight from the culture of the fast(er) to the culture of the slow(er). Chapter Three offers an overview of contemporary slow practices and theoretical approaches to slowness. Its aim is to show a variety of recent practices developing under the banner of the "slow revolution," systematize contemporary critical examinations of deceleration, and identify the key areas of critical reflection and main markers of the culture of the slow. Because of the central position of a sense of pleasure in the discourse of slowness, Chapter Four is concerned primarily with pleasure as a socio-cultural construct which, when read from the velocentric perspective, exemplifies the role speed plays in the articulation of particular cultural meanings. When approached from the perspective of alternative hedonism, a concept worked out by Kate Soper, pleasure reveals its potential to initiate processes of dearticulation and rearticulation of cultural practices promoted by the mainstream culture of high speed. In the light of alternative hedonism, slow living may be read as a form of quest for these types of pleasure that have been repressed and marginalized by accelerated consumerism.

In the third part of the book, entitled "The Other Speed: Cultural Practices and Representations," I examine the ways in which ideas of deceleration have permeated particular spheres of everyday life and the ways in which they have been represented. Chapter Five places deceleration in the context of space and interprets cultural ramifications and representations of slowing down both in private (home) and public (urban) spaces. The analysis of the latter from the perspective of deceleration allows me to extend the research area beyond the Slow Cities movement, an offshoot of the Slow Food movement, and to approach a new spatial consciousness brought to life under the influence of slow life philosophy as a manifestation of new territoriality, a term I use to refer to spatial practices that are informed by a desire to restore place as a repository of cultural meanings and to overcome the social and cultural effects of deterritorialization.

Chapter Six is concerned with slow travel and its representations in two texts: "A Slow Travel Manifesto" and the guidebook *Slow London*. While the analysis of the manifesto allows me to examine the rhetoric typical of the narratives of slowness, the analysis of the slow guide helps identify components of the slow travel experience. A change of cognitive paradigm, which consists in the shift of the focal point from the visual to the multisensory, emerges as one of the main features of the discourse of slowness and chimes with the postulates put forward by Carlo Petrini, a founding father of the Slow Food movement, advanced to start a systematic defence of our senses endangered by the homogenization of "fast" taste, a product of globalized culture.

In the last chapter of the book I examine slowness in the context of popular culture and the popularity of the so-called "relocation products," i.e. television programmes and self-help books, which cast the move from city to country as a panacea for the tensions, anxieties and nervousness of accelerated urban life. I argue in this chapter that a new literary phenomenon, which I propose to call *slow lit*, has emerged at the intersection of the culture of relocation and chick lit. Popular literature written in the conventions of chick lit about a new life in the country is indicative of a certain expansion of the narratives of slow living in the area of popular culture and of the vitality of slow ideas as well as of their marketing potential. Suspended between an urbanite's dream and utopia, the representations of life change proposed by slow lit are read as a part of the process of romanticization and mythologization of slow living.

While the primary aim of this book is to explore the cultural and social dimensions of deceleration and their representations in the context of the cultural resistance they inspire, the perspective from which this is done is marked not so much by a desire to dethrone high speed, but rather to integrate the two realms of speed. The options we have do not need to be reduced to an either/or rhetoric, such as characterizes Mirko Zardini's claim that "we are now faced with conflicting paths to salvation: to accelerate into an even-faster world, beyond the realm of physical movement, or a return to the speeds of the past."[13] Troubled himself by such a reductionist vision, Zardini revises this view and suggests a change of emphasis and reformulation of the proposed solution which signals a departure from the logic of the either/or choice. Zardini proposes a perspective which seems to be more practical and which stems from his reflections on "which models and which associations can break the tyranny of productivity upon speed."[14] The claim opens up a space for critical examination of a possibility to

---

13 Mirko Zardini, "Preface," in *Speed Limits*, ed. Jeffrey T. Schnapp (Milan: Skira, 2009), p. 22.
14 Zardini, "Preface," p. 23.

connect and integrate what modernity has disconnected. In an age of flow, globalization and homogenization, fragmentation and mobility, slowness has been increasingly vested with the values that are associated with connection to people, places and environment, and represented as a connecting principle in a disconnected world.

While slowness has emerged as a strategy of contestation and resistance, in the course of time it has undergone various changes, the main one being its shift from the margin to the mainstream of cultural practices. Slow practices have installed themselves safely in the cultural landscape of high modernity and have grown to form a recognisable lifestyle option available to man; some phenomena, such as Slow Food and Slow Cities, have become consolidated institutionally and developed identities recognizable all over the world; the ideas of slow life have entered popular culture and initiated a rise of various products which feed on the international repute of slow. The phenomenon's embeddedness in contemporary culture allows us to identify those aspects of cultural deceleration which have withstood the test of time and have proved to have a lasting appeal for contemporary man; on the other hand, it invites one to register and examine tensions and contradictions that have started to appear within it and pose questions concerning both the future of slowness and the rise of a *post-slow* stage in the history of speed.

Part One

Speed: Towards Velocentric Perspective

Chapter One: "To Fly like Mercury": Acceleration and Modern Experience

> Only against the background of speed
> can slowness be determined and learnt.
> Helga Novotny, "Time – The Longing for the Moment"
>
> Anything that is away is too far-away. [...]
> And anything that lasts, lasts simply too long.
> Wolfgang Sachs, "Speed Limits"

"The Erle of Tolous," a late medieval English chivalric romance, starts with an introductory prayer in which an author asks God to "Gyff us wele to spede/And gyff us grace so to do/That we may come thi blysse unto/On rode as thou dydest blede."[1] When read by a contemporary reader, a humble request to "Permit [us] to speed well" ("Gyff us wele to spede") may demonstrate one of the most concise versions of the evolutionary history of the term "speed." Interestingly, archaic as it might be, the phrase turns out to be emblematic of a present-day mindset since the romance uses the meaning of the word "speed" which to contemporary speed *aficionados* may have a very special appeal. Old English *spēd*, Middle English *spede* or Old German *spout* originally meant "prosperity;" the meaning of speed as "quickness of motion" emerged in late Old English thus causing the word to develop two meanings at the same time.

The analysis of the etymological history of the word "speed" reveals an interesting paradox and illustrates the change in the role speed plays in our lives today. The old-fashioned use of the word "speed" is anachronic only apparently. In fact, it is this anachronic meaning that grasps the very essence of modern understanding of the social and cultural role of speed. The etymology of the word "speed" reveals a significant semantic transfer. What used to be the first meaning of the word – to succeed, to prosper – has been lost but although the word "to speed" suggests today fast movement in time, what it connotes is success, prosperity and luck, the qualities that the word denoted in the past. In the course of time the first denotative meaning of the word evolved into a connotative meaning thus reflecting cultural significance of speed in modern

---

1 "The Erle of Tolous," in *Codex Ashmole 61: A Compilation of Popular Middle English Verse*, ed. George Shuffelton (Kalamazoo, MI.: Medieval Institute Publications, 2008), p. 83.

experience. Although today "speed" means the rate or a measure of the rate of motion or swiftness of action, the archaic meaning of the word has established itself as an inseparable connotation in a culture in which speed is the guarantee of social, economic and political success, and provides a conceptual framework for a study of the cultural dimensions of acceleration.

This chapter is devoted to an analysis of changes in attitudes towards speed that have been brought about by technological development. Absent from social consciousness until the 17th century and not recognized as a category constitutive of identity, in the last two centuries speed has become an integral element of culture's signifying practices. My aim is to analyze the evolution of the Western attitude to mechanical speed. Since mechanical speed is said first to have been the product of the railways, the analysis carried out in this chapter starts with the epoch that preceded the building of the first railways and whose character was marked by the development of the coach. In this chapter I propose to look at the history of acceleration from the perspective of how the successive launching of new technologies facilitating mobility and communication (coaches, trains and cars) affected (re)conceptualizations of speed and the role it played in the lives of people and led to its gradual domestication and naturalization in the experience of the everyday.

1. The Mercurial Start: "In a stage coach (if God permits)"[2]

The history of speed is the history of the ways we talk about it. Or the ways we keep silent about it. Mathias Rieger's reading of the thirteenth-century text by Frederick II *On the Art of Hunting with Birds* shows the Emperor as an observer whose insightfulness and sensitivity to the meaning of details, as well as his ability to establish connections between facts and see them in a wider perspective, endowed him with aptitudes required from a researcher, scientist or thinker today. Yet, in one respect, as Riegers says, Frederick II demonstrated the extent to which his mode of interpretation of the world around him was the product of the cultural conditioning of his times. "[H]e never talks about

---

[2] The phrase comes from the announcement for a stage-coach running between London and York: "Whoever is desirous of going between London and York [...] Let them Repair to the Black Swan in Holborn [...], where they will be conveyed in a Stage Coach (if God permits), which starts every Thursday at Five in the morning" as referred to in *The Book of Days: A Miscallany of Popular Antiquities in Connection with the Calendar*, ed. Robert Chambers, Vol. 2 (London: W. & R. Chambers, 1832), p. 227.

speed,"[3] writes Rieger, whose interpretation of the Emperor's descriptions of the hunting falcon exposes a fundamental difference in the ways the world is read and explained now and then.

Modern as he might sound, Frederick II produced an account of hunting which lacks what for a man today might seem to be the first, most obvious association – that of speed and racing. For the Emperor the flying speed of the falcon was not responsible for the success of its performance because, as Rieger observes, "the idea of seeing the reason for the falcons [sic] success in his extraordinary speed *could not occur to him*" [emphasis mine].[4] The intellectual frame of mind of the thirteenth-century man, regardless how incisive his approach to his subject may seem now, attracts attention of a contemporary thinker for what it does not embrace rather than for what it accentuates.[5]

For centuries, speed did not matter in the life of individuals. It did not matter in the sense that for centuries speed did not feature as a formative component of daily experience. Neither did it divide a community into the fast and the slow nor did it engender cultural practices that would incite further acceleration. It was not an element of the cognitive frame of mind of the man in the past, the fact fittingly summarized by Jeffrey T. Schnapp:

> During the tens of centuries that human beings rode horses velocity remained a cultural theme of mostly secondary significance. Either cosmic (as in the whirling of the celestial spheres) or comic (as an inversion of the *gravitas* expected of authority figures), it was either otherworldly – the attribute of angels, demons or gods – or, if this worldly, identified with the lower social orders: with messengers, couriers, entertainers, and thieves.[6]

However, the fact that for centuries the world was neither categorized nor described in terms of speed-oriented practices should not eclipse what is emphasized by Paul Virilio: that it was the speed of the development of weapons systems that marked the pace of the progress of history. For Virilio speed is the fuel of

---

3   Ivan Illich, Matthias Rieger, Sebastian Trapp, "Speed? What Speed?" in *Speed – Visions of an Accelerated Age,* eds. Jeremy Millar, Michiel Schwartz (London: The Photographers' Gallery and the Trustees of the Whitechapel Art Gallery, 1998), p. 150.
4   Illich, Rieger, Trapp, "Speed? What Speed?" p. 150.
5   The question of the role speed played in the social and cultural dimensions of premodernity has recently been examined by Jeffrey T. Schnapp. See: Jeffrey T. Schnapp, "Spinners," which is the first in the series of extracts from a book in progress entitled *Quickening (An Antropology of Speed)* to be published in *West 86th*. The excerpts are available at http://www.west86th.bgc.bard.edu/articles/spinners.html
6   Jeffrey T. Schnapp, "Crash (Speed as Engine of Individuation)," *Modernism/modernity,* Vol. 6, No.1 (1999), p. 9.

history. The implication of his claim that "history progresses at the speed of its weapons systems"[7] is that speed is the motive factor of social and cultural development and that it is closely connected with power and territoriality. The ways which facilitate a faster and more efficient disempowerment of the enemy are the guarantee of victory. According to Virilio, it was speed that marked new borders and defined new constellations of political powers.

Until the 17th century the everyday experience, which would include "[…] the performance of music, falconry and fishery" as well as "commerce, medicine, and architecture […]" among many other aspects of life, "thrived without reference to it [speed]."[8] Starting at the turn of the 17th century, technological development set in motion a spiral of increasing speed, which energetically expropriated successive spheres of life and began to affect the character of everyday life. The stage-coach, the successor of the wagon, that "cart without springs,"[9] started to change the lifestyle and landscapes of people in Europe. According to Schnapp, it was the rise of coaches, "not the development of railways or motocars" that "first set the terms for the entanglement of modern notions of subjecthood with experiences of accelerated motion."[10] Seen as "great improvements on all the then existing conveyances," the stage-coach "[…] was destined to work great changes in travelling,"[11] which would in the not-so-distant future range from making travelling more available to those less privileged, through the creation of the need to improve the quality of existing tracts and roads, many of which were impassable after periods of bad weather, to the shortening of travelling time. The changes which were brought about by the popularization of stage-coaches reached their climax in the 18th century. Following the claim made by Thomas Hylland Ericksen that the history of the last 200 years can be viewed as the history of acceleration,[12] and Wojciech Józef Burszta and Waldemar Kuligowski's analysis of the turn of the 19th century in terms of the "epoch of the first acceleration,"[13] it seems justifiable to call the 18th century the century of the running start.

---

7   Paul Virilio, *Speed and Politics*, trans. Marc Polizzotti (Los Angeles: Semiotext(e), 2006), p. 90.
8   Illich, Rieger, Trapp, "Speed? What Speed?" p. 154.
9   *The Book of Days*, p. 227.
10 Schnapp, "Crash," p. 4.
11 *The Book of Days*, p. 227.
12 Thomas Hyllard Eriksen, *Tyranny of the Moment. Fast and Slow Time in the Information Age* (London: Pluto Press, 2001), p. 51.
13 Wojciech Józef Burszta, Waldemar Kuligowski, *Sequel. Dalsze przygody kultury w globalnym świecie* (Warszawa: Muza SA, 2005), p. 52.

While still rarely used in the 16th century, stage coaches flourished in the 17th and 18th centuries, to be gradually replaced by trains only in the 19th century. Three main routes available at the end of the 17th century in Britain were used by an increasing number of travellers; moving with a speed of two or three miles per hour, the coaches were capable of covering the distance between London and Liverpool in 10–12 days depending on the season.[14] The growing awareness of the correlation between speed and economic profit was a vital motive force behind the improvement of the design of the stage-coach, a fact best illustrated by the competition between British merchants. The race begins in the mid-18th century when the merchants of Manchester decide to invest their money in the construction of an improved type of a coach: "The Flying Coach," as it was called, travelled at a speed of four miles per hour. Rivalry between Liverpool, Sheffield and Leeds merchants caused the coaches to develop "the respectable velocity of eight miles an hour,"[15] thus making English coaches perhaps the fastest in Europe at that time. Samuel Coleridge in a letter to his wife (1799) comments on the superiority of English coaches over their German counterparts: "[i]n England I used to laugh at the 'flying waggons': but compared with a German Post Coach, the metaphor is perfectly justifiable, and for the future I shall never meet a flying waggon without thinking respectfully of its speed,"[16] although the appreciation of the speed of the conveyance did not stop him from confessing that "travelling in chaises, or coaches even, for one day is sure to lay me up for a week."[17]

The serviceableness and utility of stage-coaches were not things that were equally enthusiastically welcomed by everybody. Negative emotions produced by the growing popularity of coach travelling were reflective of concerns of social, medical, psychological and economic nature. In *The Grand Concern of England Explained in Several Proposals to Parliament* (1673), stage-coaches were represented as threatening the physical condition of people: "those who travel in these coaches contracted an idle habit of body; became weary and listless when they rode a few miles, and were then unable or unwilling to travel on horseback, and not able to endure frost, snow or rain, or to lodge in the field."[18] Coaches were thought to put a traveller in a state of lethargy, which threatened his vitality

---

14 *The Book of Days*, p. 227.
15 *The Book of Days*, p. 228.
16 *Letters of Samuel Taylor Coleridge*, ed. Ernest Hartley Coleridge, Vol. 1 (Boston: Houghton, Mifflin and Company, 1895), p. 279.
17 *Letters of Samuel Taylor Coleridge*, p. 356.
18 *The Book of Days*, p. 227.

and stamina and produced a general sense of laxity; a traveller by coach would reach his destination devitalized, enervated and defeated by the hardships of travel.

In 1772 the American Quaker John Woolman walked in six weeks from London to Yorkshire; his refusal to ride a horse or use a stage-coach was motivated by disquietude caused by stories about cruel treatment of both people and animals working for the stage-coach companies. His walk was supposed to help other members of the Society of Friends, the church he belonged to, see a source of social injustice and make them react:

> Stage-coaches go upwards of one hundred miles in twenty-four hours; and I have heard Friends say in several places that it is common for horses to be killed with hard driving, and that many others are driven till they go blind. Post-boys pursue their business, each one to his stage, all night through the winter. Some boys who run long stages suffer greatly in winter nights, and at several places I have heard of their being frozen to death. So great is the hurry in this spirit of this world, that in aiming to do business quickly and to gain wealth, the creation at this day doth loudly groan.[19]

Woolman saw the increasing sense of rush as fuelled by the blind pursuit of wealth, which, in turn, was given a boost by the development of technologies of mobility. The vicious circle of speed, greed and technology was beginning to turn and make life more unnerving and destabilizing. Published in 1771, Tobias Smolett's novel *The Expedition of Humphry Clinker* created an image of the city which was the site of constant and chaotic movements and where people of different backrounds and occupations crowded, pushed and jostled against each other. The stage-coach was a vital part of this setting:

> In short, there is no distinction or subordination left –The different departments of life are jumbled together –The hod-carrier, the low mechanic, the tapster, the publican, the shopkeeper, the pettifogger, the citizen, and courtier, all tread upon the kibes of one another: actuated by the demons of profligacy and licentiousness, they are seen every where rambling, riding, rolling, rushing, justling, mixing, bouncing, cracking, and crashing in one vile ferment of stupidity and corruption – All is tumult and hurry; one would imagine they were impelled by some disorder of the brain, that will not suffer them to be at rest. The foot-passengers run along as if they were pursued by bailiffs. The porters and chairmen trot with their burthens. People, who keep their own equipages, drive through the streets at full speed. Even citizens, physicians, and apothecaries, glide in their chariots like lightening. The hackney-coachmen make their horses smoke, and the pavement shakes under them; and I have actually seen a

---

19 Robert Balmain Mowat, *Americans in England* (Boston: Houghton Mifflin Company, 1935), p. 39.

waggon pass through Piccadilly at the hand-gallop. In a word, the whole nation seems to be running out of their wits.[20]

The age of the running start – the age of coaches – witnesses the birth of the fast man, the first type of the accelerated man, whose identity from this moment on will be increasingly constructed by speed-oriented practices. For Schnapp the fact is indicative of the rise of "a new subjectivity and a new social identity" since wheels recast social divisions to such an extent that

> [n]ew class distinctions arise: between drivers and pedestrians; between owner-drivers and driver-chauffeurs; between seekers of thrill and seekers of passage; between owners of coupés, phaetons, enclosed carriages, carts and hackney cabs. Other distinctions blur as lords, clothiers, clerks and dancers merge into the undifferentiated form of mass spectacle and collective rush known as *traffic* [emphasis original].[21]

The category of the accelerated man permits identification of the moment in which the control of one's speed and mobility became the sign of one's identity. One of the most important implications of the process initiated in the last decades of the 18th century was the generation of new socio-spacial arrangements which required a new demarcation line to separate the fast time of "tumult and hurry" from the slow time, and the "fast" spaces from the "slow" spaces. In Louis Sébastian Mercier's *Tableau de Paris* (1781) we read that the author "has earned the right to criticize the barbaric luxury that are today's coaches, having been knocked to the pavement on three different occasions and nearly run over alive."[22] For the first time not being fast enough was felt to be responsible for the sense of failure. The act of *not using* already existing, available technologies of acceleration would amount to gradual relegation of a non-user to the peripheries of time, thus creating the foundations of different temporalities. In the future, the advance of the railway will push forward the process of generating different "geographies of time," to quote the title of Robert V. Levine study of culture-inflected perceptions of time, since as Eric Hobsbawm notes "[…] in widening the gap between the places accessible to the new technology and the rest, it intensified the relative backwardness of those parts of the world where horse, ox, mule, human bearer or boat still set the speed of transport."[23] The age of coaches

---

20 Tobias Smollet, *The Expedition of Humphry Clinker* (Ware, Hertfordshire: Wordsworth Classics, 1995), pp. 80–81.
21 Schnapp, "Crash," p. 14.
22 Louis Sébastian Mercier, quoted by Schnapp in "Crash," p. 14.
23 Eric John Hobsbawm, quoted by Jeremy Stein in "Reflections on Time, Time-Space Compression and Technology in the Nineteenth Century," in *Timespace. Geographies*

prepared the ground for changes to come. The division between the slow and the fast, although not yet exerting its full power, started, for the first time, to be experienced by a growing number of people.

While some tended to perceive coach travelling in terms of their negative social and somatic side-effects, for others stage coaches had a revolutionary economic potential which was not properly used. Disappointed by "the dilatoriness of coach traveling,"[24] which made communication between Bath, where he ran a theatre, and London problematic and thereby affecting his business performance, John Palmer, a British reformer, prepared a plan to improve information exchange between distant places in the country. After having to confront strong resistance from the British government, he made Britain the first country to use mail coaches and build a national postal service. The introduction of the postal system in eighteenth-century Britain impelled the improvement of roads and speeded up the process of the improvement of the design of coaches, which altogether further increased the speed of travel. By the beginning of the 19th century, a stage coach could travel at a speed of 10 miles per hour. Thomas De Quincey observed that "[t]he mail-coach it was that distributed over the face of the land, like the opening of apocalyptic vials, the heart-shaking news of Trafalgar, of Salamanca, of Vittoria, of Waterloo."[25] The process of information exchange entered a new phase.

The uniqueness of mail-coaches, noted De Quincey, consisted not only in the role they played in the process of information distribution, but also in the fact that they were the last mode of travel which did not separate man from the physical experience of speed: while travellling on a mail coach a person was not detached from the experience of speed since "we heard our speed, we saw it, we felt it as thrilling; and this speed [...] was incarnated in the fiery eyeballs of the noblest amongst brutes, in his dilated nostril, spasmodic muscles, and thunder-beating hoofs."[26] For the fast man speed was still a part of the travel experience. It was its familiarity and visibility that made traveling by coach a form of an experience almost organically linked with the natural world.

But it was the same organic union between the human and the animal that would be the source of speed limits. No matter how sophisticated and well

---

of Temporality (London: Routledge, 2001), p. 110. See also: E. J. Hobsbawm, *The Age of Capital 1848–75* (London: Weidenfeld and Nicolson, 1975), p. 60.

24 *The Book of Days*, p. 228.
25 Thomas De Quincey, *The Works of Thomas De Quincey: The English Mail Coach*, Vol. 4 (Edinburgh: Adam and Charles Black, 1862), p. 288.
26 De Quincey, *The English Mail Coach*, p. 302.

conceived the construction and design of the "flying wagons" were, the rate of movement of the traveling coach in the end would always be limited by an unimprovable factor – the natural speed of a horse. The liberation of speed from the confinement imposed by the regime of natural speed, overcoming of which had been, up till then, nothing but a fascinating dream, will be brought by a steam-powered engine. Its construction and application in everyday life will call into being the fast-forward man whose identity will be marked by a departure from the qualities of the fast man and by the conviction that whatever is fast can be made faster. De Quincey's essay may serve as a point of departure for further theorization of the difference between the fast man, the product of the age of the running start, and the fast-forward man, the product of the mechanical speed brought by the railways. The rise of fast-forward man corresponds with the development of commodity-centered culture of transportation, which is passenger-oriented and focused on "mass transit"[27] rather than on the movement of individuals.

The concept of fast-forward movement exposes a desire to speed up what seems to be unnecessarily slow and to shorten waiting time. As such, it may become a useful term with which to describe the nature of the change that the mode of thinking about speed was undergoing at the turn of the 19th century. What created the conditions for the identity of the fast-forward man to thrive was the liberation of speed from the limitations created by human physicality. The latter, seen as an obstacle on the way to annihilating time and space, was now reduced to a form of inconvenience likely to be removed, sooner or later, by the power of the human mind. The railway heralded the death of the fast man and marked the beginning of the domination of the fast-forward man for whom the disappearance of the speed horizon became a fundamental element of the new cultural setting.

2. "Catch Me Who Can": Fast-forwarding and Velocitization of Culture

While the 18th century was the century of the running start, the 19th century was a century of rapid acceleration caused by the Industrial Revolution, a time when people woke up to old dreams of unlimited speed and when new demands arouse. Horse-drawn carriages slowly became a relict of the past; engine power inaugurated the era of "the new speed."[28] Constructed in 1803 by Richard

---

27 Schnapp, "Crash," p. 3.
28 Carl Honoré, *In Praise of Slowness* (New York: Harper One, 2005), p. 23.

Trevithick, an early steam-powered road vehicle was an element of a new tale told by "the new speed" man. In the case of Trevithick the narrative line of the tale would reach its climax one year later when in February 1804 the Victorian engineer would produce the world's first steam engine to run successfully on rails. In 1808 Trevithick's attempts to construct a locomotive whose steam engine would not crack the iron rails resulted in the construction of a new locomotive named by its constructor "Catch Me Who Can."

"The rush for higher speeds is a cultural fall-out of the steam engine,"[29] observed Wolfgang Sachs. The nineteenth-century acceleration cast speed as limitless and made possible what for the centuries had been restricted to the realm of imagination – the overcoming of natural speed – which revealed a fundamental principle: when speed is experienced as having no limits, the race is on; when the race is on, the velocitization of culture starts since the increasing speed of one element enforces the increase of speed of other elements of the system. The process of velocitization of culture, which in the words of Carl Honoré, "fuels a constant need for more speed,"[30] may be illustrated by the ways in which the rail system responded to the emergence of the automobile. John Urry identified the speed response, which consisted in the "building of new rail lines that permit the running of very high-speed trains,"[31] as one of the major side-effects of the emerging car culture. According to the principle of velocitization, a fast car necessitates the advance of a faster train and thus the process of acceleration is put on a track of neverending competitiveness.

The history of steam engine is the history of velocitization. Both the speed response of the rail system and the concept of velocitization correspond with what Thomas Hylland Eriksen recognized as one of the principles of speed when he said that speed is contagious: "[i]f one gets used to speed in some areas, the desire for speed will tend to spread to new domains."[32] The result is that "constant acceleration is the one constant in our experience of modernity,"[33] the principle

---

29 Wolfgang Sachs, "Speed Limits," in *Speed – Visions of an Accelerated Age*, eds. Jeremy Millar, Michiel Schwartz (London: The Photographers' Gallery and the Trustees of the Whitechapel Art Gallery, 1998), p. 123.
30 Carl Honoré, *In Praise of Slowness. Challenging the Cult of Speed* (New York: Harper One, 2005), p. 34.
31 John Urry, *Mobilities* (Cambridge: Polity, 2007), p. 110.
32 Eriksen, *Tyranny of the Moment*, p. 71.
33 Jeremy Millar, Michiel Schwartz, "Introduction – Speed as a Vehicle," in *Speed – Visions of an Accelerated Age*, eds. Jeremy Millar, Michiel Schwartz (London: The Photographers' Gallery and the Trustees of the Whitechapel Art Gallery, 1998), p. 16.

constitutive of the identity of the fast-forward man whose hunger for speed can never be satisfied. Once the dream of unlimited speed had been released from the confines of human (and animal) physicality and the concept of natural speed suspended, speed, now controlled by man, was relocated to a sphere whose character was shaped by human competitiveness.

While coach travel was one of many "pre-industrial mobility-systems,"[34] train networks, which started to grow in the mid-19th century, were among most important mobility systems that laid the foundations of modern culture, "re-ordering the contours of time, space and everyday life."[35] The familiarity of speed, its experiential availability and proximity, were lost as a consequence of the arrival of the train, the moment in the history of cultural acceleration marked in De Quincey's words by the fact that "[b]y now, on the new system of travelling, iron tubes and boilers have disconnected man's heart from the ministers of his locomotion."[36] The difference between coach speed and train speed was described by De Quincey in terms of the difference between "a consciousness" and "a fact of our lifeless knowledge, resting upon *alien* [emphasis original] evidence."[37] The rise of railway speed, which as Małgorzata Nitka says, was "nothing more but an objectively calculated rate of motion, a mere arithmetic, a piece of second-hand information the travel is removed from"[38] brought about its defamiliarization.[39] While for a coach traveller the experience of speed was an integral element of the journey, for a train traveller speed would become "not only mechanical but also invisible, unfamiliar, and therefore unrelated and indifferent to the traveller."[40]

The defamiliarization of speed was the consequence of the fact that "[i]n the Machine Age, neither the body nor the topography any longer define a natural limit for speed."[41] Almost one hundred years after the first steam train took passengers along a public line Amédée Ozenfante was to write in *Foundations of Modern Art* (1928): "[i]f speed has a limit, it would interest us little: but speed has no limit other than that of light, and from us to that… Ideal speed will never be attained, but the impulse towards that infinite constitutes a lever that can never

---

34 Urry, *Mobilities*, p. 13.
35 Urry, *Mobilities*, p. 92.
36 De Quincey, *The English Mail Coach*, p. 303.
37 De Quincey, *The English Mail Coach*, p. 302.
38 Małgorzata Nitka, *Railway Defamiliarization. The Rise of Passengerhood in the Nineteenth Century* (Katowice: Wydawnictwo Uniwersytetu Śląskiego, 2006), p. 22.
39 Nitka, *Railway Defamiliarization*, p. 25.
40 Nitka, *Railway Defamiliarization*, p. 23.
41 Sachs, "Speed Limits," p. 123.

exert its full power."[42] The steam engine created the new imaginarium of speed by opening wide the doors to spaces which for centuries had belonged to the realm of utopian dreams.

The transition from stage-coach speed to train speed is meaningful in yet another, disturbing, respect. Written only twenty-five years after De Quincey's essay, *A Fast Life on the Modern Highway: Being a Glance Into the Railroad World from a New Point of View* by Joseph Taylor anticipates the complexity and ambiguity of man's relation to speed and technology, the qualities which in the years to come will become dominant. Taylor's journey on the engine of an express train rather than in the passengers' coaches is an illuminating experience which enforces reconceptualization of defamiliarized speed and reveals its deeply schizophrenic status, which at the time Taylor was writing was still an obscure, unverbalized suspicion rather than a fully realized warning. Suspended between the sense of the newly achieved domestication of speed and the recognition of its monstrous power lurks the germ of ambiguity. Domestication and defamiliarization, nearness and distancing, proximity and alienation will form an alliance whose integrity not only will not to be threatened but will become quite natural in the years to come:

> A train at night is a spectacle of terrible magnificence anywhere, but we have become so familiarized with it that it has lost its force, and we simply regard it in the ordinary realistic light in which we look on any other casual object. We can stand unmoved on a railway, see the iron mass that whirls a helpless freight of our fellow-creatures sixty miles an hour dash flashing past us, hear the scream and rush, feel its hot blast on our face, and the earth quivering under our feet without the slightest emotion.[43]

This excerpt is illustrative of the swift process of domestication of the railways and making speed inconspicuous. Taylor's commentary foretold what in future would constitute the very essence of the modern attitude to speed. The time needed to tame the beast of the steam engine turned out to be very short. By 1874 a steam train was no longer a source of excitement for travellers. Train speed was naturalized and started to permeate the fabric of everyday life to such an extent and in such a manner that it ceased to stand out as something either marvelous or terrifying and awesome. Taylor's words anticipated what in future would become radicalized by the omnipresence of speed: speed would be seen as

---

42 Amédée Ozenfante, quoted by Jeremy Millar in "Rejectamenta," in *Speed – Visions of an Accelerated Age*, eds. Jeremy Millar, Michiel Schwartz (London: The Photographers' Gallery and the Trustees of the Whitechapel Art Gallery, 1998), p. 104.

43 Joseph Taylor, *A Fast Life on the Modern Highway: Being a Glance Into the Railroad World from a New Point of View* (New York: Harper&Brothers Publishers, 1874), p. 96.

man's inalienable right rather than a privilege it used to be in times of restricted access to fast transportation. Moreover, in the future man would be capable of noticing speed only when deprived of it and forced to slow down. The invisibility of speed in everyday life and its naturalization, processes whose beginnings Taylor registered, reach their climax when "speed is often only made apparent by what we moderns would perceive of as its lack, that is slowness."[44] The figure of the 1874 man, unmoved in the face of technological acceleration, is emblematic of a human readiness to embrace acceleration and allow speed to penetrate every aspect of life.

Taylor's account of his experience of travelling on a locomotive reveals yet another dimension which can be described in terms of a double speed paradox. Firstly, the more energetically speed makes its way in the everyday life and the more naturalized it becomes, the more disconnected from it a person becomes; secondly, a person achieves not so much control over speed as an *illusion* of it. The aristocratic German traveller, Friedrich von Raumer, the author of the first account of the first railway journey from Liverpool to London, seemed to be blinded by his fascination with the ingenuity of the human mind when he declared that "a human being turns the monster to his will with the touch of a finger."[45] To him it was a monument to the triumph of a scientific mind capable of imposing its own order on the natural order of things.

Taylor's tale, equally expressive of fascination with "the awful power of man,"[46] resonates, nevertheless, with a disquieting intuition which makes one think that von Raumer might have been too fast to eulogize the power of man to control the monster of steam energy. When confronted with the power of the engine, Taylor realizes that "you are in other hands than your own."[47] The feeling is reinforced when Taylor happens to witness the opening of the furnace:

> In a blind fog, with the whistle screaming in my ears, the wild echoes booming and reverberating from every part of the roofed station, the hot furnace licking in the coal at my feet – I could see nothing, I could do nothing, and I held tightly onto the rail, stunned and helpless.[48]

From "stand[ing] unmoved" and "without slightest emotion" to the unnerving sense of being confronted with a power that reveals human frailty and helplessness,

---

44 Millar, Schwartz, "Introduction – Speed as a Vehicle," p. 16.
45 Friedrich von Raumer, quoted by Sachs in "Speed Limits," p. 123.
46 Taylor, *A Fast Life on the Modern Highway*, p. 96.
47 Taylor, *A Fast Life on the Modern Highway*, p. 97.
48 Taylor, *A Fast Life on the Modern Highway*, pp. 99–100.

the multifariousness of the reactions speed and technology produced is emblematic of an approach which in the years to come would grow to penetrate contemporary modes of thinking about acceleration. Taylor's account attracts readers' attention with the author's insightfulness and his sensitivity to these aspects of technological development and cultural acceleration which in future will play a decisive role in the formation of the character of social relations. The figure of the engine-driver provokes the feeling of gratitude which "we all owe to the fine, conscientious, laborious fellows into whose hands we entrust our lives."[49] The reliability of the driver, his dedication and his know-how deserve the utmost praise since these are the qualities that enable travellers to make the best use of the iron horses. Yet we need to remember that "[s]peed is fascinating because it confers power,"[50] as Sachs says, and the remark is indicative of what Taylor was only vaguely aware of, sensing it rather than being able to verbalize it openly, but what nowadays emerges as the foundation of power relations. The separation of man from speed which was defamiliarized and turned into "a fact" and "*alien* evidence," as well as the delegation of the administration of speed to the mass of anonymous and invisible engine-drivers created the demarcation line separating those with access to speed and those who were either deprived of it or whose access was severely limited, and turned speed to a measure of empowerment.

The evaluation of the impact of the railway system on the social, economic and cultural dimensions of everyday life in the 19th century habitually highlights the introduction of timetables, the regimes of the clock-time followed by the inevitable loss of social and private meanings of kairological time, the increasing valorization of punctuality as well as the compression of space and the emergence of new sociabilities, e.g. the railway compartment and the station.[51] When approached from the perspective of the evolution of the fast man, the history of the railways reveals its transformative potential with reference to the meaning of speed and its reconceptualization. What matters is not only the fact that goods and people are carried from one place to another but that they are carried *with speed*. Urry says: "[a]n incredibly powerful, speeding mechanical apparatus is foregrounded as a relatively familiar feature of everyday life."[52] Speed may

---

49 Taylor, *A Fast Life on the Modern Highway*, p. 100.
50 Sachs, "Speed Limits," p. 126.
51 For a more detailed discussion of the abovementioned aspects of the impact of the railway systems see: John Urry, *Mobilites*, Wolfgang Schivelbush, *The Railway Journey. Trains and Travels in the Nineteenth Century*, Małgorzata Nitka, *Railway Defamiliarization. The Rise of the Passengerhood in the Nineteenth Century*.
52 Urry, *Mobilities*, p. 93.

provoke a variety of responses, ranging from enthusiasm to hostility, but it remains clear that it was relocated from the realm of the imaginary to the realm of not only the real but above all, the everyday. The dream of overcoming the limitations of human physicality and harnessing technology in the service of man seemed more real then ever before. High speed is no longer a dream; it is a now a project which is to be realized in order to guarantee efficiency, quality and comfort, concepts for which speed has become a synonym in modern era.

3. "Faster!... faster still!": Car Culture and the Theology of Speed

By the beginning of the 20th century speed was already a fetish. Taylor's premonition, his sense of witnessing a power control of which might have been illusory, did not discourage subsequent waves of enthusiasm inspired by the prospects of living in an accelerated world. For Eric Hobsbawm the 20th century started in 1914 and ended in 1991.[53] In his analysis of political forces which shaped the character of the "Age of Extremes," 1914 played the role of a demarcation line behind which radical political and social transformations started to change Europe for ever. Were the history of speed and acceleration taken as an alternative point of reference for a new periodization, it might be the year 1909 that would mark the symbolic beginning of the twentieth century. It was then that Henry Ford installed his first assembly line and that F. T. Marinetti, in his famous *Manifesto of Futurism*, wrote: "[w]e affirm that the world's magnificence has been enriched by a new beauty: the beauty of speed. A racing car whose hood is adorned with great pipes, like serpents of explosive breath... a roaring car that seems to ride on grapeshot is more beautiful than the Victory of Samothrace."[54] The symbolic meaning of this date is also recognized by John Tomlinson, who sees 1909 in the context of the emergence of "unruly speed," which oscilates around "excitement, thrill, danger, risk and violence."[55]

Inspired by a desire "to free this land from its smelly gangrene of professors, archaeologists, *ciceroni* and antiquarians,"[56] Marinetti's manifesto is among the

---

[53] Eric Hobsbawm, *The Age of Extremes: The Short Twentieth Century, 1914–1991* (London: Michael Joseph, 1994). See also: T. H. Ericksen, *Tyranny of the Moment*, pp. 9–10.
[54] F.T. Marinetti, "The Founding and Manifesto of Futurism," in *Documents of 20th Century Art: Futurist Manifestoes*, ed. Apollonio Umbro, trans. Robert Brain, R. W Flint, J.C Higgitt, Caroline Tisdall (New York: The Viking Press, 1973), p. 20.
[55] John Tomlinson, *The Culture of Speed. The Coming of Immediacy* (Los Angeles: Sage, 2007), p. 44.
[56] Marinetti, "The Founding and Manifesto of Futurism," p. 22.

most explicit demonstrations of the fascination with speed and its creative potential to liberate the world from the confines of the past.[57] Marinetti, a poet and revolutionist, grasped the essence of the change which at the turn of the 20th century started to redefine the quality of everyday life and spatial and temporal experiences of every individual: "Time and Space died yesterday. We already live in the absolute, because we have created eternal, omnipresent speed."[58] His announcement of the death of time and space foreshadowed many contemporary discussions about the annihilation of space and the social and cultural consequences of redefinition of the meanings of time.

Marinetti's valorization of speed as a fundamental tool of change and the marker of modernity is followed by its aesthetization. The poem "To My Pegasus" written in 1908, originally published as "To the Automobile," as well as the 1912 poem "The Pope's Monoplane," are emblematic of Marinetti's fascination with technology and its wild, dynamic beauty. "To My Pegasus" is a product of the technological euphoria characteristic of the beginning of the 20th century. In the poem it finds its manifestation in the anthropomorfization of the car, this "handsome demon" and a "racing monster," which, nevertheless, is not man's enemy. The car's energy, speed and insatiable hunger "for horizons" permit one to "leap/ with ecstasy, into liberating Infinity!.../" The untamed energy of the beast, which provokes sexual unrest, is a source of liberating force which challenges the limitations of human corporeality:

> Faster!... faster still!...
> And no respite, and no rest!
> Release the brakes!... You can't?...
> Smash them then!...
> Let the engine's pulse centuple!
>
> Hurrah! No more contact with the filthy earth!...
> At last, I break loose and fly freely
> over the intoxicating abundance
> of Stars streaming in the great bed of the sky![59]

---

57 It is a coincidence that some eighty years later it was Italy again that became the home of a movement whose roots were closely connected with reflections on speed, i.e. the Slow Food Movement, an international movement which initiated a long process of rehabilitation of slowness.
58 Marinetti, "The Founding and Manifesto of Futurism," p. 22.
59 F. T. Marinetti, *Selected Poems and Related Prose*, ed. Luce Marinetti, trans. Elizabeth R. Napier, Barbara R. Studholme (New Haven: Yale University Press, 2002), pp. 38–39.

A similar tone of obsessive desire to escape the limitations of human physical conditions permeates "The Pope's Monoplane," a poem in which Marinetti demonstrates his fascination with the plane. The image of the "filthy earth" still propels Marinetti's imagination when he talks about "horror of the earth," the prison-like space of confinement which turns one's room's intimacy into a suffocating sense of enslavement that can only be overcome by the energy and speed of movement:

> Horror of my room like a coffin with six walls!
> Horror of the earth! Earth, dark lime
> trapping my bird feet!... Need to break free!
> Ecstasy of climbing!... My monoplane! My
> monoplane![60]

The filthier the earth becomes, the more beautiful man grows; the smaller the earth gets, the larger man becomes. In the shrinking world of new technologies of mobility man seems to himself even bigger and more important. Speed plays a significant role in the process of disarming eternal human fears and anxieties generated by the limitations of one's physicality. Capable of challenging space and time, man begins to cherish the vision of entering a realm limited so far to the gods.

Published in 1903, "A Song of Speed," this "imperishable utterance," as William Archer called it in *The New York Times*, is the poetic record by William Ernest Henley, an English poet who, crippled and housebound for many years, experienced a car drive for the first time in his life. This "daring and vivid celebration of the new factor in human affairs"[61] – speed – in the eyes of the poet is "a rupture/ An integral element/In the new scheme of Life."[62] The poem rolls like a wheel; its rhythm remains uninterrupted; the images flow as if seen from a moving car, one followed by another to create a sequence which forms a transgressive experience and places a traveller by car in a space beyond time.

Henley introduced his poem with a fragment of "Canto X" from *Don Juan* by Lord Byron, an eager and enthusiastic admirer of speed, who wrote that "nothing gives a man such spirits [...]/As going at full speed." For Byron speed was a tool of transgression: it allowed one to overcome the limitations of human

---

60 Marinetti, *Selected Poems and Related Prose*, p. 43.
61 William Archer, "Henley's A Song of Speed," *The New York Times*, 20 June 1903, accessed 4 March 2014, http://query.nytimes.com/mem/archivefree/pdf?res=F20F1 4F73D5D11738DDDA90A94DE405B838CF1D3
62 William Ernest Henley, *A Song of Speed* (London: Long Acre, 1903), p. 10.

corporality and made a man like a bird whose natural capacity for flying was the object of human perennial envy. The image of the flight is an expression of a desire for transcendence and liberation from the constraints of the human condition. The ability to produce "speed vehicles" empowered man and compensated for his obvious physical imperfection. Watching racing motorcycles in 1902 Mario Morasso analyzed his own experience of speed. The spectacle of power and energy produced "the peculiar sensation composed of admiration, trepidation and excitement" and corresponded with Henley's vision of speed, as for Morasso the experience of speed opened "some unknown boundary between reason and madness, between the possible and the impossible, between the real and the miraculous, between the human sphere and the superhuman one."[63]

For Henley speed was God's creation. By identifying speed as God's gift to man, Henley suggested that speed elevated man to the level of the divine:

> Speed!
> Speed, and the range of God's skies,
> Distances, changes, surprises;
> Speed, and the hug of God's wings
> And the play of Good's airs,
> Beautiful, whimsical, wonderful;[64]

Speed, a "shadow of Deity,"[65] allows for the appreciation of the beauty and perfection of the God-made world, and thus speeding becomes an act of religious worship. While the figure of an engine-driver was the source of admiration for Taylor, a car-driver incited similar emotions in Marcel Proust, who remembered the mechnic-driver with whom in 1907 he went to Lisieux to visit his parents. The intensity with which the driver held the stering wheel, an almost religious state of concentrations, made Proust discern a semblance between the driver and the apostles:

> Most of the time, however, he clung to a wheel [...] not unlike the cross of consecration carried by the apostles alongside columns of the choir at Sainte-Chapelle in Paris, or the cross of Saint Benoit, or, more generally, stylizations of wheel in medieval art. He was so motionless that it seemed as if he wasn't steering, but rather holding it up as a symbol of his identity, much like saints in the entryways of cathedrals hold up an anchor, a wheel, a harp, a scythe, a gridiron, a hunting horn, or brushes.[66]

---

63 Mario Morasso, "Sensations of Speed," in *Speed Limits*, ed. Jeffrey T. Schnapp (Milan: Skira, 2009), p. 228.
64 Henley, *A Song of Speed*, p. 10.
65 Henley, *A Song of Speed*, p. 20.
66 Marcel Proust, "Motoring Days," in *Speed Limits*, ed. Jeffrey T. Schnapp (Milan: Skira, 2009), p. 234.

through a corporeal experience of acceleration."[86] The rise of vehicular space created a new context for the experience of the everyday in which natural space is increasingly cast as a site of conquest. Sachs observes that "[i]n the Machine Age, neither the body nor the topography any longer define a natural limit for speed. As a consequence, the modern notion that human motion was set on an infinite path toward ever increasing acceleration could take hold in the popular mind."[87] The speed of the mechanical vehicle bridged the gap between the real and the fantastic and made speed-oriented imagination and lifestyle an integral component of the everyday. Speed started its evolution into "a cultural thematic" of modernity and a "sacrament of modern individualism."[88]

The history of the development of Western systems of transportation and communication is a fascinating story about the power of the human imagination and human mind, and the challenge of human physicality is one of the most powerful motive forces that propel development of societies. Zygmunt Bauman is right when he says that modernity "is born under the stars of acceleration and land conquest [...]."[89] However, right from the very beginning of its development, the culture of velocity has been marked by tensions, anxieties and contradictions. Sometimes openly and explicitly declared, sometimes unvoiced, they have been indicative of "the other history of speed" underwritten not so much by the ideas of progress and development as by the sense of emergence of an addiction which has placed the individual on a one-way track. As Schnapp puts it, he "finds himself caught in an addiction loop, threatened, on one hand, by monotony, and, on the other, by the need for ever new stimuli in order to maintain the same level of intensity."[90] As will be shown in the next chapter, this insatiable desire for more speed has become one of the defining practices of contemporary culture.

---

86   Esther Milne, "'The Minister of Locomotion': Some Historical Speculations on Velocity Culture," *M/C: A Journal of Media and Culture* 3, (2000), accessed 13 October 2013, http://www.api-network.com/mc/0006/ministers.php.
87   Sachs, "Speed Limits," p. 123.
88   Schnapp, "Crash," p. 10.
89   Zygmunt Bauman, quoted by Tomlinson in *The Culture of Speed*, p. 78.
90   Schnapp, "Crash," p. 4.

Chapter Two: Streamlined Culture and the Rise of the Mis-man

> But today a new form of speed permeates society [...].
> Robert Hassan, *Empires of Speed*

"Demand more from your broadband," suggests British Sky Network helpfully and makes an advertising slogan work as an emblem of the type of lifestyle which in conditions of excess, acceleration and hyper-individualism has been extensively naturalized. "Demand movies with a muffin, songs at the station and browsers at the bar. With thousands of WiFi hotspots from The Cloud Sky Broadband Unlimited lets you demand it all,"[1] asserts the ad, while sketching a portrait of a mutifunctional individual and marking the contours of what may be described as his "technologically natural" environment. Although to say that today an individual is a nomad permanently on the move is to verge dangerously on platitude, it is precisely this piece of common knowledge that reflects the character and extent of changes brought about by the development of technologies of mobility. As Andreas Schafer and David. G. Victor claim, "[t]oday, world citizens move 23 billion km in total; by 2050 that figure grows to 105 billion."[2] Though physical mobility has reached an unprecedented scale, it is not its quantative but rather qualitative dimension that requires new analytical approaches. Equipped with the props of new speed, a modern nomad is evolving into a "digital nomad"[3] who does not circumscribe his life space – he annuls boundaries by opening them wide. The car ceases to be a symbol of mobility; now it is an iPhone that allows not only mobility but also simultaneity, by bringing the world close to its user and making it stretch itself open in front of him and wait invitingly to be explored. WiFi hotspots are gates to an expanded world which is no longer limited by here and now. A journey in physical space is accompanied by a travel in virtual space; both types complement each other thus creating a new dimension of mobility in the case of which it is not so much the number of kilometers travelled but rather the number of experiences accumulated that describes mobility of a present-day traveller.

---

1 *The Times Magazine*, 22 September 2012, back cover.
2 Andreas Schafer, David G. Victor, "The Future Mobility of the World Population," *Transportation Research,* Part A 34 (2000), p. 171.
3 See: Tsuqio Makimoto, David Manners, *Digital Nomad* (Chichester: John Wiley & Sons, 1997).

The world that lies within hand's reach requires from its inhabitants multi-functionality and a readiness to embrace and negotiate between the various options it endlessly multiplies. The modern imperative to have more and faster generates a highly demanding attitude underpinned by a redefinition of availability of goods and services (speed included) in terms of an inalienable right rather than a privilege. In the context of the richness of the offer monofunctionality becomes a sign of an inability to come to terms with the multithreading of what Gilles Lipovetsky calls "hypermodernity."[4] It becomes a manifestation of one's techno-disability and impairment.

The advertisement cited earlier is illustrative of the rise of a new type of speed-oriented experience located at the intersection of Jeffrey T. Schnapp's two types of the experience of velocity.[5] Schnapp distinguishes between "thrill-based" experience, which he associates with the experience of driving a coach and a car, and "commodity-based" experience, whose essence consists not so much in the sense of rupture and ecstasy that speed generates as in the sense of its serviceability. The ad inscribes itself in the area of convergence between the two thus leading to the demarcation of the space whose character is marked by coexistence of agency and passivity. "Shielded from [...] the engine,"[6] a contemporary "passenger" is submitted to velocity (in the above case this refers to the speed of an internet connection and its availability) but it is his readiness to negotiate the terms of its usage that replaces passive reception of speed with its active appropriation. Schnapp's "kinematic subject"[7] is as much a product of acceleration as he is the producer of a wide spectrum of cultural practices. The source of thrill is of different origin here – it is not so much speed, as this is taken for granted, but rather the ability to pass successfully between various temporal and spatial regimes.

The focus of the previous chapter was on the evolution of mechanical speed, which brought about the transport revolution responsible for the breaking up of transportation's "'mimetic' relationship with natural forces"[8] and which introduced a modern man into the space so far exclusively reserved for utopian visions of the future. This chapter continues to reflect on changing shapes of relations to speed while focusing specifically on two developments which will be approached as central for contemporary speed-oriented analysis: the rise of

---

4   Gilles Lipovetsky, *Hypermodern Times* (Cambridge: Polity, 2005).
5   Jeffrey T. Schnapp, "Crash (Speed as Engine of Individuation)," *Modernism/modernity*, Vol. 6, No. 1 (1999), p. 8.
6   Schnapp, "Crash," p. 18.
7   Schnapp, "Crash," p. 18.
8   John Tomlinson, *The Culture of Speed. The Coming of Immediacy* (Los Angeles: Sage, 2007), p. 16.

streamlined culture, whose contours are marked by a constant elimination of forces that oppose the relative motion of an object, and the appearance of the third category of the accelerated man i.e. the category of the multifunctional, instantaneous and simultaneous man – the *mis-man* – who is shown as the embodiment of the evolutionary climax of the process of acceleration and the symbol of dominant conceptualizations of speed, time and mobility.

With reference to a growing number of works concerning the ways speed has been reconfiguring social relations and shaping cultural practices and also to the results of the analysis conducted so far, I propose to introduce a critical perspective which throughout the book will be referred to as *velocentric* and which helps us view speed not so much as a *product* but rather as a *producer* of culture and thus as a tool of cultural analysis. Understood as a rate of movement which may be fast or slow rather than associated only with quick i.e. high speed, speed is approached as central for its culture-generating potential. When read in the light of velocentrism culture emerges as a site whose shape is molded not only by the fast but also by the slow. The application of velocentric perspective allows for the embrace of the variety of cultural practices of "deceleration lane" whose rise will be discussed in Chapter Three and whose everyday dimensions and manifestations will be the subject of the analysis in the remaining chapters of the book.

1. "Speed in a straight line": New Conquests

To grasp the specificity of dominating conceptualizations of speed and its role in social, cultural and political life of modern societies let us go back to the beginning of the 20th century to hear F. T. Marinetti speak again. This will help us mark the direction of the development of speed, which, we are tempted to think, would have been observed by Marinetti with mixed feelings: the prophet of Italian futurism laid, quite unintentionally, the foundations for a critique of the ways in which speed came to be experienced by contemporary man.

In "The New Religion-Morality of Speed" we find a distinction Marinetti makes between two types of speed:

> In an s-shaped curve with double bends, velocity achieves its absolute beauty, for it is struggling against: (1) the resistance of the ground, (2) the various atmospheric pressures, (3) the pull of gravity formed by the empty space in the curves. Speed in a straight line is massive, crude, *unthinking*. Speed with and after a curve is velocity that has become agile, acquired *consciousness* [emphasis original].[9]

---

9   F. T. Marinetti, "The New Religion-Morality of Speed," in *Futurism. An Anthology*, eds. Lawrence Rainey, Christine Poggi, Laura Wittman (New Haven: Yale University Press, 2009), p. 228.

For Marinetti, a zealous and visionary worshipper of speed, differentiation between two types of velocity was a *sine qua non* of the growth of self-consciousness of speed. He wrote: "[w]e must continually vary our speed so that our mind is actively participating in it."[10] By no means should Marinetti be seen as a crypto-defender of the slow. Slowness meant for him "adoration of obstacles" and "idealization of tiredness and repose"[11] and was never valued for its inherent qualities. Nevertheless, slowing down helps underline the splendor of speed. The call to depart from the monotony of increasing speed reveals the inner tensions of acceleration. For Marinetti the experience of speed is burdened with aesthetic threat – the fast which does not enter into a dialectic relation with the slow deprives itself of its aesthetic dimension by losing self-consciousness. The rhythm marked by a change from fast to slow to fast again excavates and multiplies the sense of the glory of speed.

Marinetti's comment was a warning against what he saw as vulgarization of speed through the annulment of differentiation of its rates and his anxiety was brought about by a desire to protect the beauty of speed. The paradox is that while technological development in the last decades of the 20th century brought acceleration which Marinetti had never dreamt about, it was "speed in a straight line" that became an object of desire for modern man. Very soon this found an enthusiastic supporter whose ideas of urban design shaped Western cityscapes: the way in which the valorization of "speed in a straight line" was embraced and internalized by cultural practice is illustrated by the works of Charles-Edouard Jeanneret Le Corbusier.

In 1924 Le Corbusier wrote in *L'Urbanism* (*The City of To-morrow and Its Planning*): "[w]e are part of that race whose dawn is just awakening. We have confidence in this new society which will in the end arrive at a magnificent expression of its power."[12] The association of Le Corbusier's rhetoric with the one used by Marinetti justifies numerous attempts to look at their works from the perspective of similarities and shared ideas rather than to accentuate the difference between them. Tomlinson's rereading of Le Corbusier leaves behind this intellectual tradition and liberates it from the "mistake of suggesting his affinities with the Futurists."[13] Let us reread, after Tomlinson, the fragment from the first chapter of *The City of To-morrow*:

---

10 Marinetti, "The New Religion-Morality of Speed," p. 228.
11 Marinetti, "The New Religion-Morality of Speed," p. 225.
12 Le Corbusier, *The City of To-morrow and Its Planning*, trans. Frederick Etchells (London: The Architectural Press, 1971), p. 5.
13 Tomlinson, *The Culture of Speed*, p. 33.

A modern city lives by the straight line, inevitably; for the construction of buildings, sewers and tunnels, highways, pavements. The circulation of traffic demands the straight line; it is the proper thing for the heart of a city. *The curve is ruinous, difficult and dangerous; it is a paralysing thing* [emphasis mine].[14]

For Marinetti, an aesthete and poet, the vulgarity and unconsciousness of speed in a straight line can only be tamed by the curve, which exposes the beauty of speed; for Le Corbusier, a pragmatic urban planner, the curve is a blow struck at what he valued most, namely, a functionalism which Tomlinson describes as "instrumental rationality."[15]

"The curve is ruinous:" in the context of modern acceleration Le Corbusier's words reveal their transgressive potential while moving beyond the space of city planning. The curve is a decelerator, an impediment, an obstacle; it has to be overcome, annulled and disarmed. Speed in a straight line becomes a new cultural imperative and as such demands new spaces and practices. Jeffrey T. Schnapp notes:

> When mail coaches, trains, and automobiles first arose, their impact was traumatic and ecstatic; decades later, their velocities had been naturalized and normalized, absorbed within everyday routines. Speed, in turn, sought a home elsewhere: at the speedway, in the sky, between the planets, beyond the solar system, in atomic particles.[16]

"[N]aturalized," "normalized" and "absorbed," speed has been taken for granted as a natural element of contemporary structures of feeling. It takes more than a journey on a super-fast train or supersonic plane to make the whole world hold its breath in amazement. It takes Austrian pilot Felix Baumgarten, who on the 4th of October, 2012 broke the skydive record and reached the unprecedented speed of 833.9 mph (1.342 kph) jumping out of a balloon capsule placed at an altitude of 24 miles (39 kilometers), i.e. the edge of stratosphere. The sound barier was broken when Baumagarten reached the supersonic speed of Mach 1.24 thereby breaking the previous record in skydiving that belonged to American pilot Joseph Kittinger, who in 1960 skydived from the height of 19.5 miles (31.3 kilometers). Or it takes Andy Green, a former fast-jet pilot, who is now described as the fastest man on Earth for breaking the sound barrier at ground level, driving at 763 mph in Thrust SSC (a supersonic car) in the Black Rock Desert in Nevada and who plans to drive at 1.000 mph. Their achievements help

---

14 Le Corbusier, quoted by Tomlinson in *The Culture of Speed*, p. 33.
15 Tomlinson, *The Culture of Speed*, p. 34.
16 Jeffrey T. Schnapp, "Spinners," in *West 86th. A Journal of Decorative Arts, Design History and Material Culture*, accessed 11 February 2014, http://www.west86th.bgc.bard.edu/articles/spinners.html#.

reconstruct the heroic dimension of speed, which has been lost in the process of its domestication; in the age of hyper-speed it is only in such feats that speed still reveals its potential to thrill.

What marked the character of the last decades of the 20th century was that speed's quest for "new homes" resulted in the conquest of the space of communication. While the character of the 19th century was molded by the transport revolution, the 20th century came to witness the communication revolution, says Paul Virilio.[17] Tomlinson rightly observes that "mechanical speed was, until quite recently, at the core of a robust and coherent set of modern social values, practices and individual meanings,"[18] but it was the rise of new technologies of communication as well as constant modernization (i.e. acceleration) of already exisiting technologies of transport that led to the creation of the conditions which brought to life new social and cultural practices. The character of changes is not only quantative but also qualitative and as such requires the application of a separate description which registers "the end of the era of mechanical speed"[19] and the rise of the new form of speed, and embraces the impact of unprecedented transformations and its tempo on the ways in which man shapes and negotiates his entanglement with hypermodernity.

Speed uncompromisingly displays its colonising drive and appropriates new spheres of contemporary life. Once liberated from the constraints of human physicality, speed accepts no limits. The concept of speed is increasingly coupled with the concept of "conquest;" together they start to form a conceptual tandem which becomes a marker of progress, efficiency and success in an age of acceleration. While Schnapp talks about the kinematic subject establishing his empire,[20] Robert Hassan explores "two *temporal* [emphasis original] empires […]: one based upon the clock and the other upon ubiquitous information network that comprises the 'network society.'"[21] The "Empire of Speed," as he refers to the latter, constantly extends its boundaries, creates new dominions and generates practices whose novelty challenges traditional descriptions and brings reconceptualizations of already exisiting relations of culture and power as well as social competencies and activities.

Hassan postulates the use of the concept of empire with reference to speed and the ways it has been exerting its influence and shaping cultural landscape of

---

17 Paul Virilio, *Open Sky*, trans. Julie Rose (London: Verso, 1999), p. 56.
18 Tomlinson, *The Culture of Speed*, p. 89.
19 Tomlinson, *The Culture of Speed*, p. 89.
20 Schnapp, "Crash," p. 18.
21 Robert Hassan, *Empires of Speed* (Leiden: Brill, 2009), p. 2.

modernity. The claim that "[t]he Empire of Speed is an empire based upon *technological systems* [emphasis original] that are geared towards acceleration and neoliberal globalization"[22] is used by Hassan as a point of departure from which he sets off to mark the contours of two empires: the first Empire of Speed, which embraces transformations from the 18th century to the mid-20th century, and the second Empire of Speed, whose beginning he locates in the 1970s when the shape of social and political life started to be influenced heavily by "the emerging hyperspeed of social and economic acceleration."[23] What characterizes the second Empire and distinguishes it from the first Empire of Speed is, according to Hassan, the fact that "based upon computer-driven acceleration" it creates a space where "there is no one in control because politics can no longer synchronize (keep up) with the pace of change that has become an end in itself."[24]

The condition of the stability of the second Empire of Speed requires efficiency which in the context of Virilio's diagnosis of contemporary culture as the one that accepts no delay also demands flexibility. For Hassan the pairing of efficiency and flexibility and their internalization by the structures of everyday life on both microscale of an individual experience as well as the macroscale of global institutions form a distinctive feature of the second Empire of Speed. He emphasizes that the link between efficiency and flexibility in the conditions of technologically advanced culture reinforces centrality of speed:

> [...] machines, especially computer-driven machines, are efficiency, speed and flexibility *par excellance*. To substitute human agency with nonhuman computer technology is supposedly to decrease instances of inevitable human error (inefficiencies which cost money). The effect is that we are placed on a new and powerful system-logic that places vital emphasis on speed. [...] By *networking speed* [emphasis original] we network and universalize this supposed efficiency.[25]

While Hassan sees flexibility as subordinating cultural practices of the second Empire of Speed, Tomlinson, with the credit to Virilio's "law of proximity,"[26] proposes to look at what he calls the principle of immediacy as central to the contemporary analysis of speed. According to the author of *The Culture of Speed* it allows for stabilization of the analytical platform from which to approach the moment in the history of speed which is characterized by the displacement of the

---

22 Hassan, *Empires of Speed*, p. 16.
23 Hassan, *Empires of Speed*, p. 15.
24 Hassan, *Empires of Speed*, p. 17.
25 Hassan, *Empires of Speed*, p. 20.
26 Virilio, *Open Sky*, pp. 49–57.

narrative of mechanical velocity by the new narrative of speed distinguished by unprecedented "rate of delivery of experience,"[27] the quality that has become the marker of the contemporary status of speed. The analysis of the evolution of the culture of speed that he proposes reflects the sense of shift which corresponds with Schnapp's claim that speed is looking for a new home. Tomlinson's principle of immediacy

> [...] grows out of the general acceleration of practices, processes and experience associated with the institutional and technological bases of modernity and interpreted, until around the last couple of decades, by the discourses of regulated and unruly speed. But there is a shift occurring. In significant ways, and particularly as a consequence of the ubiquitous influence of telemediation, immediacy *alters the cultural terms of speed's impact, undermining some earlier presumptions and installing new commonplace realities* [emphasis mine]. In the process, the emerging condition of immediacy produces new possibilities and problems which eclipse those of an earlier era of the culture of speed.[28]

The need to introduce new analytical tools and propose new critical approaches stems from the explosion of new cultural practices which can be analyzed neither in the context of speed's relations with progress, regulation and discipline ("ruly" speed), nor in the context of speed's communication of risk and rapture, danger and pleasure ("unruly" speed). The practices whose emergence Tomlinson emphasizes as being indicative of the sense of "a shift" situate themselves outside descriptions framed by the concepts of "ruly" and "unruly" speeds and demand new critical perspectives and new theoretical approaches. His list embraces, among other things, the impact of domestic networked computers on the separation of work from leisure, the rise of the internet, the availability of everything everywhere, the growth of call centres followed by a radical transformation of business urban spaces, the development of digital photography, and speed dating and multitasking.[29] What these have in common is that they are "historically new phenomena" and "each has at least an intuitive connection with a particular mode of the acceleration of social life."[30]

While mechanical velocity came to shape and define the character of social life in the previous era, now, according to Tomlinson, it is the concept of immediacy that allows one to grasp and embrace the sense of novelty brought about by new modes of technological and cultural acceleration. The concept turns out to be useful in the process of describing the specificity of new cultural practices for

---

27   Tomlinson, *The Culture of Speed*, p. 3.
28   Tomlinson, *The Culture of Speed*, p. 10.
29   Tomlinson, *The Culture of Speed*, pp. 72–73.
30   Tomlinson, *The Culture of Speed*, p. 73.

its reference to both space and time, which Tomlinson identifies as the foundation of the cultural principle of immediacy. The concept's orientation, Tomlinson repeats citing *Oxford English Dictionary*, is on one hand, towards "freedom from intermediate agency; direct relation or connection, … proximate, nearest, close near," and on the other hand, suggests "pertaining to the time current or instant … occurring without delay or lapse in time, done at once; instant."[31] The core of the term is the absence of any sort of distortion or diversion which might be capable of forming an obstacle on the way of a movement which aspires to be smooth and uninterrupted. Immediacy rules out delays and any form of intervening medium while emphasising directness and instantaneity.

Tomlinson distinguishes three aspects of immediacy, which in the context of our analysis of the evolution of the role of speed in contemporary culture, help synthesize the character of the shift. Firstly, the principle of immediacy is useful in the explanation of the rise of a culture of instantaneity which is "accustomed to rapid delivery, ubiquitous availability and the instant gratification of desires."[32] Secondly, the principle of immediacy embraces "cultural *proximity*" which, Tomlinson argues, brings "a distinct *quality* [emphasis original] to cultural experience."[33] What the author of *The Culture of Speed* underlines is an orientation towards directness that contemporary culture displays and which is achieved by eliminating the distance between desire and its realization.[34] The idea of immediacy puts the spotlight on culture's drive towards elimination of all elements whose presence may be the source of potential delay or form an impediment to the smoothness of the uninterrupted movement. For Tomlinson "[i]mmediacy – closure of the gap – is […] the *redundancy or the abolition of the middle term* [emphasis original]."[35]

Tomlinson's analysis of this aspect of immediacy corresponds with Zygmunt Bauman's "impatience complex," emblematic of what he sees as a "new life strategy of instant joy"[36] which underlies a vast segment of cultural production. Annihilation of "in-betweenness" is manifest in all spheres of contemporary life and is framed by two equally powerful factors: an obsession with speed and a concern with the product rather than the process. From television ads which

---

31 *Oxford English Dictionary*, quoted by Tomlinson in *The Culture of Speed*, p. 74.
32 Tomlinson, *The Culture of Speed*, p. 74.
33 Tomlinson, *The Culture of Speed*, p. 74.
34 Tomlinson, *The Culture of Speed*, p. 90.
35 Tomlinson, *The Culture of Speed*, p. 91.
36 Zygmunt Bauman, *44 Letters from the Liquid Modern World* (Cambridge: Polity Press, 2010), p. 23.

promote products thought to bring immediate results – alleviation of pain or guaranteed fat reduction – to advertisements for cosmetic surgeries where before and after are miraculously deprived of the "in-between" stage, and courses which promise to teach any foreign language or to sail in a weekend, products are framed by a common belief that what matters is the achievement and not the process. This conviction strikes a chord with Tim Ingold's claim that one of the most characteristic features of Western thought is the tendency to "privilege form over process."[37]

Telemediatization, i.e. the growing dependence of the quality of everyday life on access to electronically mediated activities, the third dimension of immediacy identified by Tomlinson, is manifested in the ways in which "the media deliver experiences of immediacy to us."[38] The range of "telemediated activities," as Tomlinson calls them, includes "watching television; typing, scrolling, clicking and browsing at the computer screen; talking, texting or sending and receiving pictures on a mobile phone, tapping in PIN codes and conducting transactions on a keypad."[39] These are the activities that form the bricks and mortar of social and economic life today and create the foundations of the culture of immediacy. In an unprecedented manner and on an unprecedented scale everyday experience is shaped by media processes and technologies which demonstrate the fusion of the human and the machine. For John Urry, the appearance of industrial machines which facilitated movement and which became an integral element of human experience to such an extent that "a 'human' life cannot be *lived* without them"[40] defines modernity. It is this physical impossibility of life without a machine in the mind of someone living in the culture of speed, to paraphrase the title of one of Damien Hirst's works, that became the foundation of modern life.

Let us return once again to Marinetti, whose temper and intuition as a cultural theorist makes him, in a manner most prophetic, predict the emergence of

> the imminent and inevitable identification of man and motor, facilitating and perfecting a continual interchange of intuitions, rhythms, instincts, and metallic disciplines that are absolutely unknown to the great majority of people today and are divined by only the most clear-sighted minds. […]

---

37 Tim Ingold, *The Perceptions of the Environment* (London: Routledge, 2000), p. 198.
38 Tomlinson, *The Culture of Speed*, p. 97.
39 Tomlinson, *The Culture of Speed*, p. 94
40 John Urry, *Mobilities* (Cambridge: Polity, 2007), p. 93.

This inhuman and mechanical type, constructed for omnipresent velocity, will be naturally cruel, omniscient, and combative. He will be endowed with unexpected organs: organs adapted to the exigencies of *an environment made of continuous shocks* [emphasis mine]. Already now we can foresee an organ that will resemble a prow developing from the outward swelling of the sternum, which will be the more pronounced the better an aviator the man of the future becomes, much like the analogous development discernible in the best fliers among birds.[41]

Putting aside the moral diagnosis of Marinetti's vision, it may still be true to say that his insight into the human-machine relation reveals in a most disturbing manner elements of the contemporary relations between man and machine. For Marinetti technological achievements, which promised a passage to the world of the novelty, occupied an external position in relation to human corporeality. The car and the plane were objects of passion and desire, non-identical with the subjects who desired them. Their ontological status was different from that of a human, while what seems to be a defining feature of the modern relation of man and machine is the fact that the body of a human is in a permanent need of technological augmentation so as to be able to adapt to "an environment made of continuous shocks." Although Marinetti's imagery may look anachronic to a reader today, his insightfulness reveals itself in recognition of the inevitability of the development of new human forms. The process of corporeal changes affected by adaptation to new conditions and environments is long, yet scientists and anatomists are already capable of predicting changes in the construction of the human bone system generated by habitually repeated gestures and body movements. It is estimated that the length of the thumb and its shape are going to undergo visible change – the thumb is predicted to be longer and thinner in future – as a result of an unprecedented mobilization of that finger enforced by the writing of short text messages. "With every tool man is perfecting his own organs,"[42] wrote Sigmund Freud in 1930. Recently this optimism has been forcefully challenged by more and more frequent realizations that speed-oriented technological developments induce changes which, in some areas, make us less perfect and more dependent on the machine. In the context of the present moment of technological advance Freud's perfection means intensification of the

---

41 F. T. Marinetti, "Multiplied Man and The Reign of the Machine," in *Futurism. An Anthology*, eds. Lawrence Rainey, Christine Poggi, Laura Wittman (New Haven: Yale University Press, 2009), pp. 90–91.

42 Sigmund Freud, quoted by Anthony Synnott in *The Body Social. Symbolism, Self and Society* (London: Routledge, 1997), p. 149.

relation between man and machine, the result of which is internalization of the mechanical, which used to exist *outside* the person, and now has been relocated and placed *internally*. The media-theorist and media-activist Franco Berardi believes that

> [f]uturism exalted the machine as an external object that was visible in the city landscape, but now the machine is inside us: we are no longer obsessed with the external machine; instead, the 'info-machine' now intersects with the social nervous system, the 'bio-machine' interacts with the genetic becoming of the human organism. Digital and bio-technologies have turned the external machine of iron and steel into the internalised and recombining machine of the bio-info era.[43]

The claim that "the machine is inside us" points towards the hybridization of the human, whose body needs to be permanently augmented to face the challenges of the accelerated world, thus rendering the body anachronic rather than more perfect. The more technologically advanced the world becomes, the more "outdated" the human body grows. Without his mobile phone, car with its GPS device, and laptop with Internet connection a modern man is naked and fragile. Schnapp has no doubt that

> [...] technology actualizes what was once a poetic dream or a theological fiction. It gives rise to 'living' machines that are also machines for living: machines that, because endowed with powers of agency, intuition, and moral autonomy, are capable of serving as prosthetic enhancement of human bodies and psyches.[44]

Never before has man's body been made so incapable of keeping up with the world he himself has created, so obsolete and so old-fashioned. Urry openly states: "[h]uman powers increasingly derive from the complex *interconnections* [emphasis original] of humans with material objects, including signs, machines, technologies, texts, physical environments, animals, plants and waste products. People possess few powers which are uniquely human, while most can only be realized because of their connections with these inhuman components."[45] Marinetti's dream of "the metallization of the human body"[46] has come true although in a different setting and by means of different props. Equipped with

---

43 Franco Berardi, "Futurism and the Reversal of Future," accessed 5 March 2014, http://www.generation-online.org/p/fp_bifo8.htm.
44 Schnapp, "Crash," p. 2.
45 John Urry, *Sociology Beyond Societies. Mobilities for the Twenty-first Century* (London: Routledge, 2000), p. 14.
46 F. T. Marinetti, quoted by Tomlinson in *The Culture of Speed*, p. 56.

"contemporary electronic terminals,"⁴⁷ an inhabitant of the second Empire of Speed redefines the conditions of his presence in the physical world and the character of social relations he enters into and those he creates. The internalization of technology and the "digitalization [....] of the facilities of home and office" which are "carried around on the body or at least in a small bag making those that can afford such objects 'geographically independent'"⁴⁸ not only cause reconceptualization of boundaries between the private and the public and work and leisure but also bring about the devalorization of the cultural meaning of effort. Virilio maintains that "[…] where there is a choice between a lift or an escalator and a simple staircase to reach upper floors, no one takes the stairs. Similarly, where a metro corridor is too long and there is a moving walkway (a travolator) at the users' disposal, no one traipses along the corridor."⁴⁹ In the age of lifts, disappearing stairs become a metaphor of change whose essence consists in reconceptualization of human subjectivity, a change that today, more likely than ever, we are ready to acknowledge.

2. Speed Now: In the Culture of Disappearing Stairs

"Speed, fantastic but fatal, speed, exhilarating but frightening, fabulous but fascist,"⁵⁰ writes Jay Griffiths in *Pip Pip*. One hundred years after Marinetti published his manifesto, Wisława Szymborska, Polish poet and the Nobel Prize winner, published a collection of poetry entitled *Here*. Among the poems collected in the volume there is a one entitled "Nonreading" which in the light of what has been already said about acceleration and compression strikes one as delivering a disturbing image of a typical fast-forward man living in the world in which high speed is as natural as the air we breath. The poem, focused on what recently has become a common tendency among literate, educated members of the affluent West, reveals a transformation of the status of speed in everyday life, the character of which might have disappointed Marinetti:

> Bookstores don't provide
> a remote control for Proust,
> […]
> Seven volumes – mercy.
> Couldn't it be cut or summarized,

---

47   Tomlinson, *The Culture of Speed*, p. 105.
48   Urry, *Sociology Beyond Societies*, p. 28.
49   Virilio, *Open Sky*, p. 55.
50   Jay Griffiths, *Pip Pip* (London: Flamingo, 1999), p. viii.

> or better yet put into pictures.
> There was that series called "The Doll,"
> but my sister-in-law says that's some other P.*
>
> And by the way, who was he anyway.
> They say he wrote in bed for years on end.
> Page after page
> at a snail's pace.[51]

Seen as the side-effect of accelerated lifestyle, non-reading becomes an emblem of the change. In the conditions of turbo reality reading amounts to physical exclusion from the stream of current events and becomes a form of self-inflicted banishment from the space of mobility and flow which efficiently incapacitates other psycho-physical activities. Once we conceptualize modernity in terms of acceleration reading becomes an old-fashioned activity and as such demands, in a manner most uncompromising, exclusive attention, which in the age of mutifunctionality is a luxury.

> We live longer
> but less precisely
> and in shorter sentences.
>
> We travel faster, farther, more often,
> but bring back slides instead of memories.
> […]
> But we're still going in fifth gear
> and knock on wood, never better.[52]

The poem undermines a commonly shared technological optimism, the essence of which was grasped by Freud, who wrote: "[i]n the photographic camera he has created an instrument which retains the equally fleeting auditory ones; both are at bottom materializations of the power he possesses of recollection, his memory."[53] Szymborska's non-reader's memory has been replaced by external, digital facilites which take over the functions of human memory. The power of one's memory consists in the longevity of material objects such as slides or digital photographs.

---

* Translators' note: "The reference is to the Polish novelist Bolesław Prus (1847–1912), whose most famous work, *The Doll* (1890), later became a popular TV miniseries."
51 Wisława Szymborska, "Nonreading," *Here*, trans. Clare Cavanagh, Stanisław Barańczak (Boston: Houghton Mifflin Harcourt, 2010), p. 39.
52 Szymborska, "Nonreading," p. 39.
53 Sigmund Freud, quoted by Synnott in *The Body Social*, p. 149.

Szymborska's little poem investigates the influence of acceleration and compression of culture on an individual to declare that the quality of human life seems to be the effect of the Faustian deal: for more speed and for longer life the price to be paid is of living "less precisely and in shorter sentences." This phrase resonates with the words of Thomas Hylland Eriksen, who believes that "there are strong indications that we are about to create a kind of society where it becomes nearly impossible to think a thought that is more than a couple of inches long."[54] The shrinking and the shortening, the compression and the acceleration – these processes are behind a vast number of practices of contemporary everyday life which impose the rhythm and set the pace. While for Marinetti "[t]he joy of shifting from third to fourth gear"[55] was the joy of man entering the world of new speed, for Szymborska fourth gear is already a symbol of the technological past. Fifth gear is the symbol of human fascination with speed on one hand, but on the other hand it is the symbol of its naturalization – one does not need to dream about speed any more since it has become an integral dimension of everyday experience. Speed is no longer a challenge – it seems only natural that if something is fast, it is just a question of time before it will be even faster.

The Marinetti-Szymborska dialogic relation in the context of the gear metaphor may be seen as an illustration of what Hassan reads as a defining paradigm of contemporary speed: an open-ended form underpinned by the recognition that "the rate at which humans communicate and the rates of increase in productivity and efficiency can never be fast enough."[56] The consequences of the loss of the temporal horizon permeate every sphere of life and reinforce the conviction that what is fast can and should be made faster. When the concept of "speed limits" is both a challenge and an oxymoron, the velocitization of culture turns out to be one of its fundamental principles and entails clear reconfigurations of the status, representations and experiences of speed.

One of the most significant processes that the concept of speed has been undergoing is that of cultural internalization. We remember Joseph Taylor's train passenger, to whom we have referred in the previous chapter, "stand[ing] unmoved on a railway," not showing "the slightest emotion."[57] After a relatively short period of time mechanical speed ceased to arouse extreme emotions, evoking instead a

---

54 Thomas Hylland Eriksen, *Tyranny of the Moment. Fast and Slow Time in the Information Age* (London: Pluto Press, 2001), p. vii.
55 Marinetti, "The New Religion-Morality of Speed," p. 226.
56 Hassan, *Empires of Speed*, p. 17.
57 Joseph Taylor, *A Fast Life on the Modern Highway: Being a Glance Into the Railroad World from a New Point of View* (New York: Harper & Brothers Publishers, 1874), p. 96.

sense of the naturalness of the experience of travel in a vehicle powered by gigantic steam boilers. What in the first days of railways had been exciting and mesmerizing, and often produced a sense of awe, by 1874 was hardly visible through a thickening curtain of obviousness and the normalcy of fast movement.

One hundred and twenty four years later a passenger travelling by plane notes:

> I have flown perhaps five or six hundred miles an hour while travelling in a commercial airliner. But the now banal insight about this now banal experience is that… *there is no speed here* [emphasis mine]. A slight pressing into my seat on take-off, insulated from the engine's roar and cloaked in the unreality of carpeting and suit bags and laptops; a brief, fierce application of brakes and a reversal of engines on landing, especially if the airport is old and the runways are short. But otherwise, obviously, nothing.[58]

What the two texts have in common is the representation of the invisibilization and banalization of speed in everyday experience. Jeremy Millar and Michiel Schwarz say: "[s]peed is all around us; we can feel its effects even if we are unable to see it. Speed is both forceful and immaterial, like the turbulence from a moving vehicle, like the thrust of a jet, like a good idea."[59] A similar conviction informed the claim made by a nineteenth-century engineer, David Bandelari: "[t]here is nothing frightening about speed when the traveller who benefits from it remains unaware of his velocity."[60] Invisibilization of speed is the effect of its normalization which consists in its "regulation, rationalization, and standardized measurements."[61]

The peculiarity of these processes consists in the fact that although speed tends to be less and less conspicuous and ever more experienced as natural, it turns out to be one of the major factors marking the character of life in modern culture. Neither invisibility nor banalization mean disempowerment; on the contrary, the more invisible and inconspicuous speed becomes, the more powerful effects it exerts. The claim that "there is no speed here" is emblematic of further radicalization of the processes of the invisibilization and banalization of speed. Nevertheless, the paradox is that banal and invisible as it is, speed has acquired the status of a "cultural dominant" and has become a factor that

---

58 Mark Kingwell, "Fast Forward," in *Speed – Visions of an Accelerated Age*, eds. Jeremy Millar, Michiel Schwartz (London: The Photographers' Gallery and the Trustees of the Whitechapel Art Gallery, 1998), p. 141.
59 Jeremy Millar, Michiel Schwarz, "Introduction – Speed is a Vehicle," in *Speed – Visions of an Accelerated Age*, eds. Jeremy Millar, Michiel Schwartz (London: The Photographers' Gallery and the Trustees of the Whitechapel Art Gallery, 1998), p. 16.
60 David Bandelari, quoted by Schnapp in "Crash," p. 25.
61 Schnapp, "Crash," p. 25.

powerfully determines and shapes the quality of people's lives. Speed "determines one's destiny, more precisely, more forcefully and more thoroughly than any genetic sequence identified by the Human Genome Project ever could,"[62] while also causing polarization of the world into two central categories: those who have access to speed and those who have not. The world is divided into the Fast and the Slow, and it is one's access to speed that determines which world one inhabits. Taylor's engine driver was the prototype of speed administrators whose power has been grasped by the president of The World Economic Forum, Klaus Schwab, who declared: "[w]e are moving from a world in which the big eat the small to one in which the fast eat the slow."[63] Power lies in the hands of those who can control and monitor speed and acceleration.

The naturalization, internalization and banalization of speed in contemporary experience are the result of the pursuit of "speed in a straight line" rather than "speed with and after a curve" which, as we have already shown, is oriented towards annulment of all forms of resistance, and which creates the foundations of what may be described as streamlined culture. "The circulation of traffic demands the straight line,"[64] announced Le Corbusier. In the conditions of streamlined culture anything that causes deceleration is perceived as an impediment which needs to be removed.

Reading speed in a straight line as the symbol of streamlined culture allows us to locate the way contemporary speed-oriented culture has been reconceptualizing the hierarchy of objects of desire centrally in the foreground. For Tomlinson the main difference between the culture of mechanical speed and the culture of immediacy consists in "the conjuring away of *effort* [emphasis original] – and particularly of concerted effort – from the imagination of the achievable good life."[65] The evolution from the "effortful speed" of the early days of the transport revolution to the "effortless mediated delivery"[66] of the culture of new speed is the story of changing expectations and demands which in the conditions of immediacy revolve around the annulment of distance between a desire and its fulfillment.

Streamlined culture, new speed's new home, requires from its inhabitants the formation of a wide range of adaptive skills. Their character calls for the

---

62   Guy Redden, Sean Aylward Smith, "Editorial: 'Speed,'" *M/C: A Journal of Media and Culture* 3.3 (2000), accessed 13 February 2013, http://www.api-network.com/mc/0006/edit.php.
63   Carl Honoré, *In Praise of Slowness* (New York: Harper One, 2005), pp. 3–4.
64   Le Corbusier, quoted by Tomlinson in *The Culture of Speed*, p. 33.
65   Tomlinson, *The Culture of Speed*, p. 80.
66   Tomlinson, *The Culture of Speed*, p. 81.

introduction of a third category of the accelerated man which helps to synthesize the ways in which cultural practices and experience in the most recent decades have been redefined by the changing tempo and rhythm of everyday life. What may be seen as an unnecessary multiplication of analytical concepts is framed by the conviction that the rise of the slow man, who negotiates the terms of his presence in hyper-speed culture, requires the formation of a dialogic perspective which allows him to be placed in the context of the evolution of the accelerated man and speed relations as well as the resistance these relations have generated. Promoted by streamlined culture, a new type of the accelerated man is the *mis-man*, who in order to succeed in the conditions of prioritization of efficiency and fluidity, has to be multifunctional, instantaneous and simultaneous, and whose identity is the answer to "[t]he pressures of speed – flexibility, efficiency, multitasking."[67]

The mis-man is a symbol of the ultimate defeat of space and time, which are seen as "basic forms of hindrance."[68] The invention of machines which made corporal mobility possible and which turned the far and distant into the close and near has produced a world which has shrunk as fast as Alice's room shrunk after she drank from the magic bottle. The ability to move "Faster!... faster still!..." glorified by Marinetti brought the effect of compression of both time and space, thus enforcing a change in the ways they have been represented. The effects of the annihilation of space and of deterritorialization, which consists in the separation of space from place, have been reinforced by the process of globalization, which "is itself tantamount to a particular form of acceleration, which reduces the importance of distance, frequently obliterating it altogether."[69] To navigate successfully in this compressed environment, the mis-man has to accept the shortening of his time horizon, which annuls both past and future, and which brings him to "the point where the present is all there is [...]."[70] A useful metaphor to mark, speed-wise, the transition from the fast-forward man to the mis-man is the variously interpreted collapse of the World Trade Center towers. Read in the context of the evolutionary history of speed and acceleration, the destruction of the Twin Towers tells the story of the invincibility of Western capitalism and the power of the financial markets rather than its failure, for

---

67 Wendy Parkins, Geoffrey Craig, *Slow Living* (Oxford: Berg, 2006), p. 64.
68 Wolfgang Sachs, "Speed Limits," in *Speed – Visions of an Accelerated Age*, eds. Jeremy Millar, Michiel Schwartz (London: The Photographers' Gallery and the Trustees of the Whitechapel Art Gallery, 1998), p. 124.
69 Eriksen, *Tyranny of the Moment*, p. 51.
70 David Harvey, quoted by Urry in *Sociology Beyond Societies*, pp. 124–125.

paradoxically the Twin Towers, destroyed in the terrorist attack, were symbols of the age that has just ended – an age in which profit had to be territorial, grounded, located, and situated in physical space. Their destruction marked symbolically the coming of an age in which "[f]inancial capital is pure speed and pure, immaterial profit."[71]

Susan George, in her analysis of the history of Indo-European civilization from the perspective of social speed, distinguishes three main castes: Agriculturalists, Warriors and Priests. Farmers were the slowest since they were attached to the soil, but warriors and priests were much more mobile, the main difference between them being that the warriors' mobility was performed in space, while priests were capable of performing acts of "symbolic speed." This is how she explains the relation between divinity and speed: "[c]ommunication with the gods has to be instantaneous. Part of being divine is having the gift of ubiquity. To be in several places at once is the epitome of speed."[72] In market capitalism, George observes, the role of the priests has been taken over by financial capital:

> Industrial capital can be removed from one place and reinvested in another, but not instantaneously. It, too, is grounded. Marx told us that Money must go via Commodities to make a profit, accumulate and become Money again (M→C→M→ etc,). No such constraints apply to Financial Capital which, like the priests of old, enjoys instantaneous communication with the divine, or in this case, with profit.[73]

Seen in the light of George's analysis, the collapse of the World Trade Center is symbolic of the climax in the evolution of the mis-man, whose activity is made independent of the requirements of physical location. The mis-man is ubiquitous and thus fills the void created by absent Gods and the collapse of the religious authorities. If financial capital is pure speed, as George puts it, the mis-man performs the role of the priest communicating with the gods.

Equipped with portable, personalized terminals,[74] the mis-man makes his way in "the social wilderness [which] has been eroded by omnipresent connectivity."[75] The dictatorship of the information society preserves the individual in

---

71 Susan George, "Fast Castes," in *Speed – Visions of an Accelerated Age*, eds. Jeremy Millar, Michiel Schwartz (London: The Photographers' Gallery and the Trustees of the Whitechapel Art Gallery, 1998), p. 116.
72 George, "Fast Castes," p. 116.
73 George, "Fast Castes," p. 116.
74 Tomlinson, *The Culture of Speed*, p. 105.
75 Dalton Conley, "Cell Phones Weights Down Backpack of Self-Discovery," accessed 30 March 2014, http://www.bloomberg.com/news/2011-08-30/cell-phone-burdens-backpack-of-self-discovery-commentary-by-dalton-conley.html

a state of permanent hunger for information, the satisfaction of which is seen as a fundamental condition of being up-to-date. The permanence of the state is reinforced by the fact that it is the excess of information that marks the rhythm of information processing, thus generating social and cultural practices focused on constant updating and which require multifunctionality and simultaneity from anyone wishing to be successful in the culture of speed and excess.

Relating an example found on the Internet of a teenage girl who was said to be sending about three thousand text messages monthly, or one hundred daily, one every ten minutes, Zygmunt Bauman analyzes the impact of such a model of social practice on the character of her life and concludes that it is very likely that she lives her life, thinks, cries and laughs in the constant company of other people. What Bauman sees as "crowded solitude"[76] corresponds with an experiment conducted by the American sociologist Susan Maushart, the effects of which she described in *The Winter of Our Disconnect*. Maushart explains that the idea for the experiment was born as a result of many factors, one of them being the moment when one day she entered the room of her daughter to be greeted by her daughter's friend watching her through a webcam via Skype. The incident provoked Maushart to cut off her family (her three children and herself) from access to all technological facilities, including computers, mobile phones and television, and to set out on a "six-month screen-free adventure."[77]

Both cases, Bauman's teenage Internet addict and Maushart's experiment, provide an insight into the social functioning of the mis-man whose contours are marked by an integral element of his social identity, i.e. being an open access person whose "private time has become one which is always connectable, available, and public."[78] Internet shopping in one part of the world, a chat with a friend living in another part of the globe, a teleconference in between – these are some of most popular forms of present-day mobility which create much more than an illusion of the presence. An open access person, a new form of public man, is available to an increasing number of people, with whom he forms everyday interactions by means of a type of presence which is much more than a quasi-presence, as some tend to describe it, but still something less than a "traditional" presence which requires not only attendance and company but also

---

76   Bauman, *44 Letters*…, p. 6.
77   Susan Maushart, *The Winter of Our Disconnect. How One Family Pulled the Plug on Their Technology and Lived to Tell/Text/Tweet the Tale* (London: Profile Books, 2011), p. 1.
78   Helga Novotny, *Time. The Modern and Postmodern Experience*, trans. Neville Plaice (Cambridge: Polity Press, 1994), p. 31.

physical location in *one* place. On the level of individual experience, the state of permanent availability eradicates the sense of being alone: "that sublime condition in which one can 'gather thoughts,' ponder, reflect, create – and so, in the last account, give meaning and substance to communication."[79] One effect of the changes in the ways an open access man functions in his private space is the waning of private, holiday time, a consequence of which is the generation of conditions in which multitasking can thrive. Writes Hassan: "[…] it is a culture that increasingly makes no distinction between work and leisure, private and public, day and night, physical and virtual," and the life it produces "is a life where one's whole subjectivity blends into a flow of blurring and accelerating tasks."[80]

New conditions and new forms of social interaction redefine a whole spectrum of everyday practices and routines. It is the meaning, the sense and the experience of time that has been most radically redefined by the processes of cultural acceleration. The "tyranny of the moment," so convincingly described by Eriksen, produces millions of time slaves who, living in a culture of excess, suffer from deprivation. Time has become the object of luxury and desire when

> […] the unhindered and massive flow of information in our time is about to fill all the gaps, leading as a consequence to a situation where everything threatens to become a hysterical series of saturated moments, without a 'before' and 'after', 'here' and 'there' to separate them. Indeed, even the 'here and now' is threatened since the next moment comes so quickly that it becomes difficult to live in the present. We live with our gaze firmly fixed on a point about two seconds into the future. The consequences of this extreme hurriedness are overwhelming: both the past and the future as mental categories are threatened by the tyranny of the moment.[81]

A large number of activities one is involved in take place in a time the experience of which is beyond our capabilities and which is described by Jeremy Rifkin as "computime"[82] characterized by an instantaneous availability of information. While Eriksen talks about the tyranny of the moment and a "hysterical series of saturated moments," Urry finds the concept of instantaneous time best for the description of the contemporary dimension of the human relation to time. The inhabitants of the second Empire of Speed see their relation with time defined by the rise of communication technologies operating in line with the process which we have seen at work already, i.e. the invisibilization of speed.

---

79 Bauman, *44 Letters…*, p. 9.
80 Hassan, *Empires of Speed*, p. 23.
81 Eriksen, *Tyranny of the Moment*, pp. 2–3.
82 Jeremy Rifkin, quoted by Urry in *Sociology Beyond Societies*, p. 126.

Urry's definition of instantaneous time as characterising new technologies "based upon inconceivably brief instants which are wholly beyond human consciousness"[83] brings to mind the experience of a nineteenth-century traveller who was waiting for a train to come and a contemporary airline passenger indifferent to speed at which he is flying. The relocation of speed from the space where it could be felt and touched to the space "beyond human consciousness" brings about a constant increase in the demand for a higher and higher "rate of delivery of experience" the answer to which is the replacement of linearity with simultaneity, a condition *sine qua non* for any form of socially acknowledged success in the age of hyper-acceleration.

Since it is not the aim of this analysis to multiply oversimplified interpretations of the fast/slow opposition of black-and-white rhetoric which tends to demonize the fast and valorize the slow, what needs to be explained is that the term "mis-man" has been proposed for two reasons. Firstly, the category of the "mis-man" helps us grasp the specificity of speed-oriented lifestyle changes, and its acronymic version includes the most characteristic features of the inhabitant of the second Empire of Speed whose experience is shaped by what Hassan described as a "new form of speed."[84] Secondly, since the term "mis-man" contains an implicit critique of accelerated lifestyle, the emergence of a wide range of cultural practices centered on the idea of deceleration can be inscribed in the process of cultural resistance to processes and mechanisms embodied by the mis-man. In contrast to previously introduced categories of the accelerated man, i.e. the fast man and the fast-forward man, the category of the multifunctional, instantaneous and simultaneous man, when used in its acronymic form, ceases to be neutral and instead becomes a form of critical interpretation of contemporary condition. Connotations with poor, mistaken, wrong performance correspond to a wave of more and more distinctive criticism of modern accelerated lifestyles and cultural practices they generate. Eriksen's "tyranny of the moment" is expressive of the increasing sense of dissatisfaction with the side effects of cultural acceleration and disillusionment with what it brought.

The dystopian character of the dream that came true, which the category of the mis-man reflects, displays a particularly disturbing dimension when seen from the perspective of pro-environmental consciousness. For Wolfgang Sachs, acceleration inaugurated by the rise of the railways must not be seen in isolation from ecologically-oriented contexts, since without the use of Earth's natural

---

83 Urry, *Sociology Beyond Societies*, p. 126.
84 Hassan, *Empires of Speed*, p. 17.

resources "the mobilisation of time and space"[85] would not be possible. The ecological awakening generated by the ecological crisis allows us to see acceleration from the perspective of the effects it has on the environment. Interpreted in this context, the ecological crisis turns out to be "a clash of different time scales: the time scale of modernity collides with the time scale which governs life and the earth."[86] To illustrate the environmental costs of the accelerated lifestyle Sachs refers to exploitation of non-renewable resources and comes to the conclusion that

> industrial time is surely at odds with geological time. It is probably not exaggerated to say that the time gained through fuel-driven acceleration is in reality time transferred from the time stock accumulated in fossil reserves to the engines of our vehicles.[87]

The battle to overcome man's eternal enemies, time and space, which engaged man's creative effort creates many victims. Nature with its rhythms, too slow from the perspective of industrial time, is victimized on an unprecedented scale:

> Animals are kept in appalling conditions, diseases spread, pollutions advances, soils degenerate, the diversity of species is narrowed down, and evolution is not given enough time to adapt. A host of ecological problems in the area of agriculture derives from the fact that the rhythms of nature have been taken hostage by the high-speed economy of our time.[88]

Sachs challenges the optimistic conviction shared by many that the development of the internet and cyberspace communication will result in a less exploitative policy towards nature and its resources. He points out the alarming fact that although the emission of pollution during the transfer of data is not an issue in the age of cyber culture, "electronic equipment is environmentally more expensive than usually assumed." The production of the hardware is by no means Earth-friendly. He says: "[..] no less than 15–19 tons of energy and materials – calculated over the entire life-cycle – are consumed by the fabrication of one computer."[89]

When read in the context of Sachs's interpretation of the relation of accelaration and ecological crisis, the category of the mis-man highlights the role speed plays in the destruction of the natural environment and the destabilization of eco-systems. The space of ecological consciousness and its connection with

---

85   Sachs, "Speed Limits," p. 124.
86   Sachs, "Speed Limits," p. 126.
87   Sachs, "Speed Limits," p. 126.
88   Sachs, "Speed Limits," p. 126.
89   Sachs, "Speed Limits," p. 128.

speed may be interpreted as emblematic of our present conceptualizations of speed. Contemporary culture has come to understand speed as a dominant whose hegemony is not to be questioned; side-effects of cultural acceleration are to be recognized as inevitable in the process of the quest for what has been cast as natural: more speed.

Conclusions

Speed, velocity, rate, acceleration, deceleration, motion, distance: these are some of the terms which form the core of any analysis which recognizes speed – of transportation, communication, production, consumption or any other form of social interaction – as central. Though in contemporary cultural and social discourses the terms "speed" and "velocity" are used interchangeably (and this is how they are used in the course of this book), it is the meaning of velocity in physics that has served as a point of departure to propose what will be referred to throughout the book as *velocentric perspective.*

In physics distinction between speed and velocity, with speed being a scalar quantity and velocity a vector quantity, is crucial. While speed is defined as the rate at which distance is covered and is computed according to the formula:

$$\text{speed} = \frac{\text{distance}}{\text{time}},$$

velocity is defined as the rate at which the position of an object is changed and is computed using the formula:

$$\text{velocity} = \frac{\text{displacement}}{\text{time}},$$

where displacement denotes the change in position. Following physics' definition of velocity, which explains it as direction-oriented (in contrast to speed being indifferent to direction), a velocentric perspective places its focus on speed in the context of change and direction. If we assume that velocity is the position change per time ratio, then the very concept of velocity, when applied in a metaphoric way, may serve as an analytical tool with which to examine the character of social, cultural and political transformations caused by Western culture's drive to see speed as the "sacrament of modern individualism."[90]

In the contemporary culture of hyper-acceleration and in the popular usage the culture promotes and popularizes, speed means very fast movement and connotes high speed – not *any* speed. A velocentric approach allows for the

---

90   Schnapp, "Crash," p. 10.

creation of such a vantage point from which speed, *any* speed, is analyzed for its culture-forming potential. Speed is approached as a "general, calculable, *relative* [emphasis original] rate of movement or incident"[91] rather than rapidity and the quality of being fast only. Such an approach allows us to embrace the concepts of *low speed* and *deceleration* as reservoirs of cultural practices and routines. While, as Tomlinson puts it, "it is without doubt the increase of speed that has set the cultural agenda of modernity,"[92] the decrease of speed, when viewed from the velocentric perspective, displays a potential to rewrite the conditions of the agenda.

The centrality of speed – *any* speed – in a velocentric perspective is informed by the claim made by Millar and Schwarz in the "Introduction" to *Speed – Visions of an Accelerated Culture*. When they say that "[i]t is desire that is central to any consideration of speed"[93] they make us rethink the role that speed plays in the development of Western culture. Millar and Schwarz reverse a traditional mode of thinking about the relation between progress and speed, which casts acceleration as a side-effect of technological advance, and instead approach a desire for speed as a motive factor of technological progress. Such a reconceptualization of the causal relation, which puts a dream about speed and overcoming the limitation of human physicality at the very heart of technological transformations, invites reconsideration of human development and allows them to claim that "speed is not so much a product of our own culture as our culture is a product of speed."[94]

An analysis carried out from the perspective of velocentrism leads to identification of various research areas. The theory of social life or, as Hassan refers to it, theory of temporality which analyzes *why* speed has achieved the role of the most significant factor shaping the character of social, political and cultural life, the study of *the effects* of acceleration, the coexistence of *different* temporalities ("a Europe of two speeds," the temporal aspects of ageing) or the study of the *resistance* that acceleration has inspired, may be approached as research areas which, while remaining in dialogic relations with one another, allow for creation of foundations for a systematic and multidimensional study of speed, culture and society. The exchange between the fast and the slow, rather than ghettoization of different temporal zones and their representation as separate and immune to external influences, is given substance by a velocentric

---

91   Tomlinson, *The Culture of Speed*, p. 2.
92   Tomlinson, *The Culture of Speed*, p. 2.
93   Millar, Schwarz, "Introduction – Speed is a Vehicle," p. 16.
94   Millar, Schwarz, "Introduction – Speed is a Vehicle," p. 16.

perspective which recognizes speed's forcefulness to shape social relations and practices.

A velocentric perspective centralizes speed-induced changes and reads speed as a culture-forming factor. Seen in this light the chronological divisions discussed so far – the 18th century as the century of the running start, the 19th century as the century of acceleration, and the 20th century as the century of speed – turn out to be "epicycles within one large orbit: The Speed Revolution."[95] According to Helga Novotny it is not so much what speed (understood as high speed) facilitates and mobilizes but rather what it incapacitates that should become the subject of study and the issue of concern.[96] In the chapters that follow I seek to examine the ways in which *low speed* (or *the other speed*) has come to redefine the terms of the agenda set by acceleration. The orbit of the Speed Revolution has recently grown larger to embrace one more epicycle – that of slowness considered as a strategy of resistance, a prop of sustainable living and a dream.

---

95 Helga Novotny, quoted by Millar and Schwarz, "Introduction – Speed is a Vehicle," p. 16.
96 Helga Novotny, quoted by Millar and Schwarz, "Introduction – Speed is a Vehicle," p. 21.

Part Two
Slow Time and Alternative Hedonism

Chapter Three: "To Sit like Jove": Slowness and Everyday Resistance

> How we spend our days is how we spend our lives.
> Annie Dillard, *The Pilgrim at Tinker Creek*
>
> [A]n excess of speed turns into repose.
> Roland Barthes, *Mythologies*

"Speed," one of the chapters in Thomas Hyllard Ericksen's *Tyranny of the Moment*, opens with an excerpt of a poem by a Norwegian poet and performer, Odd Börretzen, who says: "[t]he Japanese/have reduced the time to four minutes and fifteen seconds/in a recording of Beethoven's fifth/symphony. That's how fast things are."[1] In June 1955 Glenn Gould, the Canadian pianist, recorded *Goldberg Variations* by Johann Sebastian Bach. The recording lasted 38 minutes and 27 seconds. Twenty-six years later, in May 1981, Gould entered the studio to record the piece again. This time the recording lasted 51 minutes and 15 seconds. The musician explained the reason for this change: "[…] as I've grown older I find many performances, […] just *too fast for comfort*. I guess part of the explanation is that all music that really interests me […] is contrapuntal music […] and I think […] that with really complex contrapuntal textures one does need a certain *deliberation*, a certain *deliberate-ness*"[2] [emphasis mine]. Although a general tendency is to minimize, shorten and compress almost everything in order to extend the range of possibilities and options available for immediate consumption, Gould's re-recording of Bach's music may be read not only as an expression of the musician's artistic integrity but also of his critical evaluation of the consequences of being "too fast."[3]

In the light of reflections about the nature of speed and acceleration, F. T. Marinetti's s-turn, the "s-shaped curve,"[4] which, he believed, stimulates speed's

---

1 Thomas Hyllard Eriksen, *Tyranny of the Moment. Fast and Slow Time in the Information Age* (London: Pluto Press, 2001), p. 49.
2 Glenn Gould, quoted by Michael Stegemann in "A Kind of Autumnal Repose." Liner notes. *The Glenn Gould Edition. Goldberg Variations.* CD. (Sony Classical, 1982). p. 9.
3 See: Marzena Kubisz, "Ekologia codzienności w czasach nadmiaru. Poszukiwacze skradzionego czasu," in *Znaki, tropy, mgławice* (Katowice: Wydawnictwo Uniwersytetu Śląskiego, 2009).
4 F. T. Marinetti, "The New Religion-Morality of Speed," in *Futurism. An Anthology*, eds. Lawrence Rainey, Christine Poggi, Laura Wittman (New Haven: Yale University Press, 2009), p. 228.

self-awareness by differentiation of the rates of movement, unexpectedly acquires a new dimension. His suggestion that the fast needs the slow to display its beauty may help highlight the values of the co-existence of different temporalities. This co-existence, in turn, may lead to a more sustainable development of individuals and their environment. The valorization of slowness postulated by such an interpretation of the s-turn does not advert a withdrawal from the world of speed, a Luddite-like negation of its fundamental principles or a utopian desire to create a world of pre-industrial pastoral idyll. On the contrary, the image of the s-shaped curve implies that "we must continually vary our speed"[5] and it stresses the interdependency of different paces of life and their mutual complementarity rather than a hostile juxtaposition of the fast as the enemy of the slow.

The focus in this chapter is on practical and theoretical implications of deceleration as a lifestyle and as a cultural phenomenon. With regard to a modern sense of instability and, in the words of Zygmunt Bauman, the liquidity of traditional forms of social conduct on one hand,[6] and the widely celebrated understanding of identity as a project which individuals may constantly improve and revise on the other, slowing down is represented by its practitioners as an alternative which allows one to counteract side-effects of consumerist lifestyles and/or engage in the process of self-actualization through negotiations of contemporary temporality. This chapter offers an overview of current practices whose rise and forms have been informed by critical examinations of the consequences of acceleration and which attest the recent growth in popularity of slowness. The chapter also seeks to create a systematic outline of the main analytical works that contextualize the ideological embeddedness of deceleration and to identify its inner tensions and contradictions, which are illustrative of another stage in the history of man's relation with speed.

1. "But a different need is spreading"[7]: From Speedsters to Slowniks

Although throughout the history of acceleration many individual voices expressing speed-induced anxieties have been heard, it has been in the last four decades

---

5 Marinetti, "The New Religion-Morality of Speed," p. 228.
6 See: Zygmunt Bauman, *Liquid Modernity* (Cambridge: Polity Press, 2000) and *Liquid Times. Living in an Age of Uncertainty* (Cambridge: Polity Press, 2007).
7 "Cittaslow International Charter," Attachement "A" to the Charter, p. 20, accessed 26 April 2014, http://www.cittaslow.org/download/DocumentiUfficiali/Charter_20.06.11.pdf

that criticism of the accelerated rhythm of life has entered a new stage, whose character is marked by an unprecedented valorization of slowness. The novelty consists not so much in a focus on the sense of exhaustion and disappointment brought on by the increased pace of modern life, but on the extent to which these anxieties have started to permeate various spheres of personal and social performance, and on their power to generate an emotional response among speedsters. The slogans calling for deceleration have never been uttered with such immediacy and urgency and never before have they been filled with such concern for the side-effects of acceleration. Never before have certain slow practices received such institutionally provided support and institutionalized shapes (movements, networks, convivia, communities, websites, magazines, etc.).

Contemporary literature on how to practise slowness is rich and diverse. It includes a wide range of popular guidebooks, self-help books and lifestyle magazines and a vast number of websites which popularize the idea of slowing down in almost all walks of life. Organizations such as the Association for the Deceleration of Time or The Slow City network, which share a belief that speed has become an *issue*, have become a recognisable element of the cultural entourage of the present. What is symptomatic of the position from which the "slowniks"[8] voice the character of their relations with speed is that they often see it in the context of a certain ambivalence which entails reformulations and redefinitions of practices internalized and naturalized by Western society. "But where on earth does the time saved by advanced technology vanish?"[9] asks Shin'ichi Tsuji (a.k.a. Oiwa Keibo). The question is an expression of anxiety generated by the experience of extreme time compression and disappointment caused by awareness of the double-edgedness of technological development, and points to the need to initiate changes which may lead to a (re)construction of a sense of control over time.

The ambivalence we are talking about is expressed in the title of Jeffrey T. Schnapp's anthology of speed-focused texts, *Speed Limits*. Interestingly ambiguous, the title seems to suggest the more and more frequently diagnosed schizophrenic nature of speed. Since the accelerated speed of modern life demands rapid transformation of all types of cultural practices, the title may be read as a manifestation of the human desire to go faster and faster and to explore the limits of one's physical and intellectual potential. A speed limit is thus a challenge

---

8   Harry Eyres, "London's First Slow Down Festival," accessed 10 January 2014, http://www.ft.com/
9   Shin'ichi Tsuji, "Slow is Beautiful: Culture as Slowness," in *Speed Limits*, ed. Jeffrey T. Schnapp (Milan: Skira, 2009), p. 304.

which demands confrontation. This reading of speed has its roots in the conviction that speed is a sign of progress, prosperity and mobility but what lurks disturbingly behind this self-imposing interpretation of the meaning of speed is the likelihood that ostentatious appeals to speed and acceleration may blind "velocimaniacs"[10] to their faults. It is recognized by a growing number of theorists and practitioners of slowness that there is a price to be paid for the acceleration of everything and that speed brings not only liberation but also various forms of enslavement and limitation.

Slowness is now enjoying a moment of glory. Not so long ago the immediate associations with slowness were those of backwardness, impediment to efficiency, and inability to keep pace with the modern world. As speed has become man's drug and obsession, in the popular mind anything slow was thought to be old-fashioned and to deserve marginalization, if not elimination, from the space of experience. Enda Duffy traces the origin of this aversion to slowness back to the first years of the 20th century, which witnessed a rise in new forms of mobility and an awakening of new fears and anxieties, with the fears of slowness and immobility permeating social practices and representations of modernity.[11] The echo of this repugnance could still be heard in 1988 when Pierre Sansot, a French philosopher and anthropologist, noted that

> [s]low creatures don't have a good reputation. They are deemed awkward. People claim they are clumsy, even when they perform delicate operations. They are considered oafish, even when they move about with a certain grace. They are suspected of not investing much energy into their efforts.[12]

Today, although not so much time has passed since Sansot wrote "On the Proper Use of Slowness," his words may already sound, at least in some respects, anachronistic. Even though it is true to say that speed has become a fundamental element of everyday experience and affects almost every sphere of modern life, from space and landscape organization to transportation, communication, production and tourism, it is also true that speed has generated

> the rise of distinctively modern *regimes of slowness* [emphasis mine] that encompass everything from techniques for the capture of otherwise unseen phenomena to dreams

---

10 Andrew Jackson Davis, "Mercurial Brainism of the Present Epoch," in *Speed Limits*, ed. Jeffrey T. Schnapp (Milan: Skira, 2009), p. 215.
11 Enda Duffy, *The Speed Handbook. Velocity, Pleasure, Modernism* (Durham: Duke University Press, 2009), p. 66.
12 Pierre Sansot, "On the Proper Use of Slownes," in *Speed Limits*, ed. Jeffrey T. Schnapp, trans. Christy Wampole, Jeffrey T. Schnapp (Milan: Skira, 2009), p. 297.

of escape from the 'rat race' of everyday life to forms of outright protest in the name of quality (slow food), sustainability (green development), and renewed community (new urbanism).[13]

The idea of slowing down in order to reclaim what has been lost as a result of the frenetic pace of life has turned out to have strong appeal for an increasing number of people and to have inspired both organized movements and individual practices. "Slow economy, slow technology, slow science, slow food, slow design, slow bodies, slow love.... This kind of wordplay may hold the potential to liberate our imagination. It directs our attention towards alternatives at odds with the dominant common sense of modern society [...],"[14] Shin'tchi Tsuji has observed. Proposals to slow down permeate various areas of social, cultural and economic life and form a discourse which reshapes existing practices and motivates the emergence of new, alternative ones.

Recently, the term "slowness" has acquired new connotations, released by an explosion of interest in the potential of deceleration brought into the cultural circuit by the Slow Food movement, which popularized the idea of slowing down in the international arena. In the wake of 1986 demonstrations against the opening of a McDonald's near the Spanish Steps in Rome, in November 1987 a left-wing Italian newspaper published in its wine and food supplement, *Gambero Rosso*, a manifesto which identified and exposed the connection between food, pleasure, politics and slow time. Two years later, in December 1989, when its shorter version was presented at the first international Slow Food Congress in Paris, it almost immediately became a point of reference for all those who were dissatisfied with the quality of accelerated lifestyle. Written by the poet Franco Portinari, the Slow Food Manifesto expressed fears, anxieties and concerns of a growing number of people all over the world. For the delegates from 15 countries who signed the Manifesto and founded the Slow Food movement, slowness was a crying response to the oppressiveness of the culture of speed. It was represented as an instrument of change to confront the effects of accelerated globalization, inspire new forms of counteraction against the domination of consumer culture and open up new spaces, whose character has been defined by the coexistence of tradition and modernity. The movement appealed to people with different backgrounds and different occupations, living in different parts of the world. The idea of a revolution through taste and a

---

13   Jeffrey T. Schnapp, "Fast (Slow) Modern," in *Speed Limits*, ed. Jeffrey T. Schnapp (Milan: Skira, 2009), p. 29.
14   Shin'tchi Tsuji, "Slow is Beautiful: Culture as Slowness," p. 301.

defence of the pleasures of slow time and slow food cut across the professional and economic affiliations of the enthusiasts of the new movement and demonstrated a potential to rally them around the slow cause. The applicability of slow postulates and their versatility produced emotional responses not only among farmers and food producers but also among environmentalists, social activists and all those who have experienced the negative consequences of acceleration in their daily routines.

Today Slow Food is an organized international movement with members in over 160 countries. Since 1990 the movement has had its own publishing house, Slow Food Editore, which publishes, among other things, the internationally distributed magazine *Slow*, which explores the issues of slowness from various perspectives, and also offers much more specific publications such as *Slow Fish* and *Slow Wine*. The members of the movement meet at international congresses and participate in a variety of projects launched by the Slow Food, among which are Terra Madre, an international meeting of food communities which brings together farmers and food producers, the University of Gastronomic Sciences, and Salone del Gusto, the world's largest regional food fair.

While major slow movements such as Slow Food and its outgrowth, the Slow City movement, have been rooted in a sense of dissatisfaction with dominant models of life and driven by the will to implement wider social and cultural change, many slow practices have emerged in the wake of the popularity of the idea of slowing down as an attempt to refresh or redefine existing practices by creating their "slow versions." A brief overview of the recently emerging slow practices indicates the culture-forming potential of deceleration: slow fashion promotes designs and patterns which go beyond passing trends and one-season novelties and advocates simple living, which consists in, among other things, avoidance of excessive purchases – it is an answer to the needs of customers who want to be fashionable but yet do not want to yield to the demands of being permanently up-to-date with the latest fashion and be drawn into a vicious circle of shopping;[15] slow blogging challenges the pressures of immediate commentaries and celebrates silence; slow reading encourages the pleasures of slow contemplation of texts; Slow Art Day promotes the idea of a selective contemplation of works of art; and the idea of slow leisure has revived an interest in craft skills and turned sewing, knitting and cooking into trendy pastimes.

---

15 See: Juliet B. Schor, "From Fast Fashion to Connected Consumption: Slowing Down the Spending Treadmill," in *Culture of the Slow. Social Deceleration in an Accelerated World*, ed. Nick Osbaldiston (London: Palgrave Macmillan, 2013).

An interesting example of an attempt to transplant the ideas of slowing down to an environment which by definition is hostile to slowness is found in the car culture. Slow driving revives the appeals of driving a (slow) car, which consists in the fact that the driver has to return to driving skills which have been eliminated from modern ways of driving. The paradox is that in an age of speed and acceleration, which constantly reconfigures human senses and extends them through technology,[16] the term "slow" may take on new connotations and come to represent adventure. Slow driving is associated with "almost-pioneering" days of driving when the separation between the human and the machine was still clearly marked and when driving required full physical and sensual engagement. The enthusiasts of slow driving see it as radical and extreme because it is not a mediated process in which the driver is just the controller of onboard electronics but becomes a mode of confrontation with one's own competences. The often nostalgically romanticized air of an old-fashioned mode of driving owes much to the recent rehabilitation of slowness and its representations in the contexts of authenticity of experience and the contestation of dominant models of cultural practices.

The turn towards "regimes of slowness," which reveals a desire to live more balanced and sustainable lives, has also led to the popularization of downshifting, a phenomenon which is often associated with voluntary simplicity,[17] economizing, simple living, and minimalism, practices which "revolve around the idea of voluntary choice to simplify individuals' lives in order to create a simpler,

---

16   Mike Featherstone, "Introduction," in *Automobilities*, eds. Mike Featherstone, Nigel Thrift, John Urry (London: Sage, 2005), p. 9.
17   One cannot ignore that simplicity is a "relative matter" (Richard Gregg, *The Value of Voluntary Simplicity* (Auckland: The Floating Press, 2009), p. 2). Voluntary simplicity is a conscious choice made as a result of a redefiniton of one's life goals and should be contrasted with involuntary simplicity, which is often the effect of a changed economic status and unexpected financial cuts caused by layoffs and unemployment. The ambiguity and relativity of the concept of simplicity often generate fierce criticism which allows its critics to view voluntary simplicity as yet another toy in the hands of the bored and overspent middle classes. The critics of the phenomenon emphasize that what is simplicity for affluent middle classes is often an unreachable luxury for many social groups in many parts of the world. True as it is, the fact should not obscure the whole spectrum of cultural, economic and social meanings of this quest for simplicity in contemporary culture and undermine cultural significance of the phenomenon which is gathering impetus not only in the United States but also in Australia, Great Britain and other Western countries.

healthier, and more balanced lifestyle."[18] In the 1990s the Trends Research Institute in Rhinebeck, New York, headed by Gerard Celeste, identified downshifting as one of the most significant changes in contemporary lifestyle.[19] The urban landscape of the 80s, which was densely populated by self-confident and acquisitive yuppies whose career-oriented lives were full of competition-generated adrenaline, has been replaced by one marked by the transformation of the yuppie into a downshifter at the turn of the 21st century. While for a yuppie the security of material status, measured by immediate access to luxury goods and services, was the main aim, downshifters engage in a "quiet personal revolt against the dominant culture of getting and spending."[20] Downshifting is often represented as an act of cultural and social subversion because it openly contests the premises of money-and commodity-oriented lifestyles. Downshifters challenge the requirements of consumer culture by choosing to earn and buy less and by cultivating a "simpler lifestyle with more time to do more of the things one wants to do."[21] They perceive time in terms of an object of luxury and reprioritize their lives in order to regain control over their time.

Whether downshifting should be read as a form of deceleration remains problematic because sometimes it is represented as a "nostalgic retreat to an imagined community or pastoral golden age" and a form of withdrawal from the present, while slow living tends to be marked by a "greater degree of imbrication in contemporary everyday life."[22] One might note, however, that although "ascetic frugality,"[23] a vital component of simple living philosophy, is not necessarily what the advocates of slowness seek, there are aspects of downshifting which are consistent with the ideas and activities of slow living. Angela T. Ragusa in her analysis of the correspondence between slow living and simple living observes that "[a]lthough each espouses different ideology, both encourage introspection, believe in individuals' capacity to change their lives, have philosophical roots

---

18   Franco Gandolfi, "The Downshifting Phenomenon," in *Downshifting. A Theoretical and Practical Approach to Living a Simple Life*, eds. Franco Gandolfi, Hélène Cherrier (Hyderabad: The Icfai University Press, 2008), p. 4.
19   Elaine St. James, *Simplicity. Easy Ways to Simplify and Enrich Your Life* (London: Thorsons, 1997), p. xvii.
20   Carey Goldberg, "Choosing the Joys of a Simplified Life," *The New York Times*, September 21, 1995.
21   *World Wide Words*, accessed 5 February 2014, http://www.worldwidewords.org/turnsofphrase/tp-dow1.htm
22   Wendy Parkins, Geoffrey Craig, *Slow Living* (Oxford: Berg, 2006), p. 3.
23   Parkins, Craig, *Slow Living*, p. 3.

grounded in individualism and advocate the emancipating capacity of seeking alternative existences."[24] Both movements, triggered by a desire to construct alternative modes of living, are responses to the increasing commodification and commercialization of life. While traditional social movements depended on collectivity, downshifting and slow living, which Ragusa classifies as "new social movements," are grounded in "individual identity rather than in the collective action,"[25] thus suggesting the potential of the everyday to become a space of negotiation of one's individual engagement with the present.

A variety of slow practices available today were popularized by Carl Honoré, a Canadian journalist, who is now one of the main promoters of slow life in popular culture. A quick look at the table of contents of a book which has become a popular introduction to the culture of the slow, *In Praise of Slowness* (2004), gives a sense of the ambitious scope of the deceleration programme, as successive chapters deal with Food, Cities, Mind/Body, Medicine, Sex, Work, Leisure and Children.[26] Honoré's account of his own slow experience is the passionate and enthusiastic confession of a neophyte. Having experienced a moment of illumination in Rome airport at the sight of a newspaper advertisement for *The One-Minute Bedtime Story* for busy parents, Honoré, the fast man himself, decided to refashion his lifestyle and started, with the eagerness and determination of a new-born slow man, a quest for a less hectic and nerve-wracking rhythm of life and a campaign to promote "what musicians call the *tempo giusto* – the right speed."[27]

The popularity of Honoré's writings and his campaigns, which promote the idea of slow living, show clearly that he managed to seduce those who feel distressed and overburdened by the tempo of modern life and for whom his ideas reflect their own dilemmas and concerns. It is also symptomatic of a change in recent reconceptualizations and representations of the meanings of time. The change in question signals a social need to guarantee an individual access to different temporalities and render negotiation between them possible. His popularity exemplifies the fact that a "space is being carved out in public discourse [...]

---

24 Angela T. Ragusa, "Downshifting or Conspicuous Consumption? A Sociological Examination of Treechange as a Manifestation of Slow Culture," in *Culture of the Slow. Social Deceleration in an Accelerated World*, ed. Nick Osbaldiston (London: Palgrave Macmillan, 2013), p. 113.
25 Ragusa, "Downshifting or Conspicuous Consumption?...," p. 114.
26 Carl Honoré, *In Praise of Slowness. Challenging the Cult of Speed* (New York: Harper One, 2004).
27 Honoré, *In Praise of Slowness*, p. 15.

to explore alternatives to a multitasking, conspicuous consumption version of daily existence."[28] The appeal of an alternative lifestyle advertized by Honoré owes much to the rhetoric he uses (not only in his book but also in numerous essays and in the interviews he frequently gives for popular media), which offers his readers a simplified, black-and-white image of the world represented as a site of the "battle against the culture of speed."[29] Honoré says: "[f]ast is busy, controlling, aggressive, hurried, analytical, stressed, superficial, impatient, active, quantity-over-quality. Slow is the opposite: calm, careful, receptive, still, intuitive, unhurried, patient, reflective, quality-over-quantity."[30] The rhetorical vilification of speed results from him drawing a line between fast and slow and casting the former in the role of the enemy. The either/or rhetoric, dismissive of the possibility to negotiate between the demonized fast and the angelic slow, is an important element of the strategy of resistance he proposes. The bipolarity of speed representations seems to be intended to overcome a sense of helplessness, often experienced by victims of the tyranny of the moment, by creating a polarized and often simplified version of reality, identifying the sources of anxiety and offering a solution available to everyone, regardless of his financial status.

The rhetoric of slow discourse emphasizes the revolutionary character of slowing down. Honoré is optimistic when he says that "[t]he Slow Revolution will take time, but it is coming. Even Jeremy Clarkson knows it,"[31] referring to a recommendation to drive slowly in times of economic slump made by the popular British journalist (what Honoré seems to deliberately ignore is that Clarkson's encouragement to drive slowly has not been brought about by a sense of frustration caused by the excessive speed of life but by the economic need to lower the costs of car maintenance. Jeremy Clarkson, a world famous speed addict, is very unlikely to drive in the deceleration lane). The same "revolutionary" rhetoric, explicitly articulated by Honoré, permeates the language of a brief introduction to an exhibition entitled *Taking Time: Craft and Slow Revolution*, and perpetuates the representation of cultural deceleration as an instrument of change:

> The Slow Revolution is sweeping through our fast-forward culture as people everywhere discover that decelerating helps them work, play and live better. *Taking Time* exhibition shows how craft fits into this *slow culture-quake* [emphasis mine]. It offers a thrilling

---

28 Parkins, Craig, *Slow Living*, p. 123.
29 Honoré, *In Praise of Slowness*, p. 91.
30 Honoré, *In Praise of Slowness*, p. 14.
31 Carl Honoré, "Recession? The Perfect Time to Slow Down," *The Guardian*, 24 July 2008.

reminder that every object has a story behind it and that the art of making matters hugely to all of us.[32]

Cecile Andrews, the co-founder of Seattle's Phinney Ecovillage, a neighbourhood-based sustainable community, and the author of *Slow is Beautiful. New Visions of Community, Leisure and Joie de Vivre* (2006), seems to be following the path marked out by Honoré when she writes about her experience of working with the Take Back Your Time Movement: "[t]o my mind, this is the revolution of our age – one that ties in with every other movement, from the gap between the rich and the rest of us to the future of the human race."[33] Her conviction is framed by a growing realization, shared by an increasing number of people today, that our relation to time and the way we experience it in the conditions of the culture of speed and immediacy creates foundations in which the quality of our living and our environmental wellbeing are firmly embedded. For Andrews, who is much more community-oriented than Honoré, slow living is a "catalyst of social change" which is "subversive" and "countercultural"[34] and it is always underpinned by the desire to implement changes not only on an individual but also on a social level. Her understanding of the transformative power of slow lifestyles is deeply rooted in the belief that personal change cannot be enacted without social change and thus a turn towards slowness must inevitably bring about changes of social structures and practices.

The analysis of historical representations of slowness in terms of its revolutionary potential reveals an interesting pattern. Although it is true to say that since the Mercurial principle of rush and efficiency has long defined the character of modern experience and slowness was stigmatized as a paragon of backwardness, it is also true to say that its resistive and revolutionary potential has occasionally been highlighted, although never on such a scale as today. In 1929 Paul Morand, in his work *On Speed* (*De la vitesse*), referred to an exchange of comments between himself and Henry de Montherlant, who asked: "[w]ill not a day not [sic] come when, due to the ordinariness of speed and the universal competition to which it is subject, slowness will seem the most natural way to express refinement?"[35] Montherlant's slowness, seen as an instrument of

---

32 "Taking Time: Craft and The Slow Revolution," Project Summary, accessed 12 January 2014, http://www.craftspace.co.uk/page.asp?fn=2&id=57
33 Cecile Andrews, *The Slow is Beautiful. New Visions of Community, Leisure and Joie de Vivre* (Gabriola Island: New Society Publishers, 2006), p. ix.
34 Andrews, *The Slow is Beautiful*, pp. 205–217.
35 Paul Morand, *On Speed* (fragment), in *Speed Limits*, ed. Jeffrey T. Schnapp, trans. Jeffrey T. Schnapp (Milan: Skira, 2009), p. 268.

refinement, corresponds with Honoré's slowness as a "shorthand for a better way of doing everything."[36] The rhetorical similarities do not end here. Morand, in full agreement with Montherlant, observed the rise of new "[…] trends in thought that have found a new home – Christian Science, Yogism or Vedanthism among the Anglo-Saxons, neo-Shintoism among the Japanese, and even neo-Thomism among the French" which he though to be "in opposition to this belief regarding the Sovereignty of Speed.[37] For Honoré, "[a] genuinely Slow world implies nothing less than a lifestyle revolution;"[38] Morand claimed that "[m]any other symptoms of revolt are apparent"[39] and by revolt he meant a growing sense of scepticism towards the benefits of speed. Morand's statement corresponds interestingly with contemporary representations of the revolutionary power of slowing down and allows one to place it in the context of continuity and change in the ways Western culture has been imagining slowness. Whenever slowness is valorized, the rhetoric does not seem to change. Morand's call to "formulate a new law of resistance to speed"[40] sounds familiar today; his claim that we should be "masters in setting our own pace"[41] echoes many postulates of the advocates of slow living who promote the idea of constant negotiation between various temporalities. The valorization of slowness, regardless whether made in 1929 or 1989, seems to be often motivated by a rejection of the oppressive nature of reality and the belief that a balance between the fast and the slow is an indispensable condition of, to use a term favoured by slow narratives, sustainable development of individuals and their environment.

2. Theorizing the Slow: An Overview

In the last chapter of *Tyranny of the Moment. Fast and Slow Time in the Information Age*, called "The Pleasures of Slow Time," Thomas Hylland Eriksen announced that in an age of accelerated progress slowness is in danger and suggested the creation of an integrated system, the aim of which would be to protect slowness through a network of national organizations cooperating to provide slowness with the support it needs.[42] He postulated a balanced relationship of fast and slow time

---

36 Honoré, "Recession? The Perfect Time to Slow Down."
37 Morand, *On Speed*, p. 268.
38 Honoré, *In Praise of Slowness*, p. 17.
39 Morand, *On Speed*, p. 268.
40 Morand, *On Speed*, p. 269.
41 Morand, *On Speed*, p. 269.
42 Eriksen, *Tyranny of the Moment*, pp. 154–160.

which he saw as an indispensable condition for the fundaments of "a world which is spacious enough to give room for a wide, inclusive both-and (as opposed to that Protestant principle either-or)."[43] Eriksen's contribution to the ongoing discussion about slowing down is illustrative of a gradual expansion of academic space devoted to the issue of deceleration and its cultural significance, which exemplifies a crisis of the confidence with which modern culture conceptualized speed as the finest achievement of Western creativity. While an analysis of cultural conditions at the turn of the 20th century cannot aspire to comprehensiveness without an examination of the role acceleration played in the (re)formations of social structures and practices, many recent analyses of cultural conditions at the turn of the 21st century recognize social significance of slowing down and scrutinize its various aspects.

Eriksen places his valorization of slowness in the context of a critical reading of acceleration and declares that "[d]elays are blessings in disguise," and that "[t]he logic of the wood cabin deserves to be globalised."[44] Significantly, John Tomlinson ended *The Culture of Speed* with a chapter entitled "Deceleration?"[45] His questioning tone is representative of a more insightful and multidimensional reading of the cultural ramifications of deceleration and their contradictions, and of his scepticism towards deceleration as an instrument of social change. He voices doubt in the very opening of the chapter when he talks about "the so-called 'slow movement'" and insists on keeping the square quotes "conceptually in place"[46] dropping them in the course of the chapter only for the sake of style.

Tomlinson understands the term "slow movement" to embrace both organized forms such as the Slow Food and Slow City movements and a number of "initiatives,"[47] as he calls them, which popularize deceleration in a less institutionalized manner. Both types are objects of his study in terms of their status as social movements and their potential to transform social structures and institutions, an aspect identified as a defining trait of a social movement by Manuel Castells in *The Power of Identity*, whose perspective Tomlinson adopts in his analysis. A comparison of the slow movement to other social movements, especially those which marked the social and political scene in the 1970s in Europe and the Unites States, makes Tomlinson question the slow movement's potential

---

43 Eriksen, *Tyranny of the Moment*, p. 164.
44 Eriksen, *Tyranny of the Moment*, p. 157.
45 John Tomlinson, *The Culture of Speed. The Coming of Immediacy* (Los Angeles: Sage, 2007), p. 146.
46 Tomlinson, *The Culture of Speed*, p. 146.
47 Tomlinson, *The Culture of Speed*, p. 148.

to effect a transformation of existing value-systems. He goes on to argue that irrespective of the stress both the Slow Food and the Slow City movements put on alternative and counter-resistive dimensions of the practices they engender and the concerns they articulate, they "have found their niche within the material and cultural economy of western modernity,"[48] a fact which for Tomlinson seems to discredit the quality of their ideological engagement in critical evaluations of the globalized present. Tomlinson claims that Slow Food is an example of an organization which is highly commercial and benefits from a system of links between industry and market while Cittaslow "represents the interests of a particular spatial-cultural constituency and a related localized form of capital," and he sees both movements as "defending enclaves of interest, rather than offering plausible models for more general social transformations."[49]

Tomlinson refuses to see the "slow initiatives" in terms of a social movement because of what he interprets as an absence of a sense of cultural identity constructed around slow practices and a lack of what usually sets the direction of actions for any social movement which aims to implement change, i.e. "a sharply defined 'adversary.'"[50] However, although Tomlinson is sceptical whether deceleration can be seen as a force capable of overthrowing the superiority of the fast in the culture of speed, he believes that the phenomenon of deceleration should not be dismissed from the spectrum of academic reflection because it is "a response to a complex and value-ambiguous cultural condition."[51] This is the perspective that has informed *Slow Living* by Wendy Parkins and Geoffrey Craig, the first book-length academic examination of Slow Food.[52] While Tomlinson finds the critique of globalized forces formulated by the slow movement "vague and politically muted,"[53] Parkins and Craig focus on ideological engagements of slow living which they see as a form of negotiation of the politics of temporality in conditions of an accelerated capitalism, which becomes "as much an attitude

---

48   Tomlinson, *The Culture of Speed*, p. 147.
49   Tomlinson, *The Culture of Speed*, p. 147.
50   Tomlinson, *The Culture of Speed*, p. 148.
51   Tomlinson, *The Culture of Speed*, p. 148.
52   In Notes to "Deceleration?" Tomlinson writes: "[…] if there is a comprehensive critical academic exploration of the slow movement, to my shame I have missed it." Since the year of publication of Parkins and Craig's *Slow Living* was 2006, it is very likely that it was not a case of "missing it" but rather that Tomlinson might not have been able to familiarize himself with Parkins and Craig's analysis soon enough to include it in his own study of "slow movements," which was published in 2007.
53   Tomlinson, *The Culture of Speed*, p. 148.

or disposition as an action."[54] Although Tomlinson and the authors of *Slow Living* differ in their approach to the transformative potential of cultural deceleration, they all see it as a form of response to the specificity of a cultural moment.

Parkins and Craig see Slow Food as a movement which has engaged in a process of wider social change by declaring the power of pleasure. They place their study of slow living in the context of the meanings of the everyday, globalization and the slow arts of the self. Paul Willis's observation, made in the wake of a turn towards the significance of the everyday brought about by the rise of cultural studies in the 1960s, that "[i]t is one of the fundamental paradoxes of our social life that when we are at our most natural, our most *everyday* [emphasis original], we are also at our most cultural; that when we are in roles that look the most obvious and given, we are actually in roles that are constructed, learned and far from inevitable,"[55] may be read as an illustration of the framework within which Parkins and Craig set their analysis of slow living. Having read the concept of the everyday in the context of a tradition of critical examinations established by, among others, Ben Highmore, Rita Felski, Agnes Heller, Henri Lefebvre and David Chaney, they interpret it as a site of rewriting of the narratives of the self. An approach which stresses the "creative and ethical potential"[56] of the everyday allows Parkins and Craig to read slow living through the prism of individualization, a concept introduced by Ulrich Beck and Elizabeth Beck, who claim that "[a]s the range of options widens and the necessity of deciding between them grows, so too does the need for individually performed actions, for adjustment, coordination, integration."[57] The everyday is thus represented as a space of reflexive questioning of social practices and their dynamics, which often leads to the rise of alternative lifestyles. For Parkins and Craig, the case of slow living may be seen as an illustration of the way in which "new forms of being-in-the-world, or new ethics of living are shaped from the everyday, from its messiness and fragility."[58] The uncertainty of everyday life, which has been burdened with unprecedented tensions and anxieties, may release creative approaches to one's condition and activate practices of resistance against the logic of the culture of speed.

---

54 Parkins, Craig, *Slow Living*, p. 139.
55 Paul Willis, "Shop Floor Culture, Masculinity and the Wage Form," in *Working Class Culture: Studies in History and Theory*, eds. John Clarke, Chas Critcher, Richard Johnson (London: Hutchinson, 1979), p. 185.
56 Parkins, Craig, *Slow Living*, p. 7.
57 Ulrich Beck, Elizabeth Beck-Gernsheim, quoted by Parkins, Craig in *Slow Living*, p. 7.
58 Parkins, Craig, *Slow Living*, p. 8.

The imperative to participate actively in the processes of individualization enacted in daily life acquires new clarity in the context of globalization, which, as Wojciech Burszta and Waldemar Kuligowski remind us, is itself a form of acceleration.[59] Parkins and Craig emphasize a link between contemporary models of everyday experience and aspects of globalized development, and read slow living as embracing practices and narratives which are a form of response to globalization. Their approach to cultural deceleration is informed by a conviction that global processes, which give shape to social, economic, cultural and political structures and organizations, should not be viewed only in terms of their negative influence on individually and collectively experienced sense of identity and difference. Their analytical perspective, framed by an emphasis on the exchange between the global and the local, suggests interpreting of the Slow Food movement as a form of production and consumption which challenges the processes of economic and cultural homogenization. With reference to placed identities, which are often interpreted, in the words of Philip Crang and Peter Jackson, as "being obliterated" and "made less authentic"[60] by the processes of globalization, Parkins and Craig stress that although often causing a sense of departure from traditional social structures and values, globalization at the same time opens up spaces for active negotiation of the consequence of detraditionalization by the revalorization of the local, which becomes "a site of resistance in global culture."[61]

In a world in which social meanings of traditions and rituals have decreased and human identity is increasingly represented in terms of a project, the everyday becomes a site of meaningful choices made by an individual to "adjust, coordinate and integrate" present-day tensions and concerns and to navigate between "high-consequence risks."[62] As the sense of stability traditionally provided by social frameworks tends to disappear from a spectrum of modern experience, the choice of a lifestyle has become a condition for the creation of a coherent narrative of the self. "It is only with a plurality of options that we are

---

59   Wojciech Józef Burszta, Waldemar Kuligowski, *Sequel. Dalsze przygody kultury w globalnym świecie* (Warszawa: Muza S.A., 2005), p. 48.
60   Philip Crang, Peter Jackson, "Geographies of Consumption," in *British Cultural Studies*, eds. David Morley, Kevin Robins (Oxford: Oxford University Press, 2001), p. 330.
61   Parkins, Craig, *Slow Living*, p. 10.
62   Anthony Giddens, *Modernity and Self-Identity. Self and Society in the Late Modern Age* (Stanford, CA.: Stanford University Press, 1991), p. 4.

forced into a reflexive stance towards all our options,"[63] notes David Chaney in his analysis of lifestyles as markers of contemporary life. Parkins and Craig see modern culture as offering a chance to engage in "self-artistry"[64] by providing us with tools which are helpful not only to search for modes of self-expression and self-realization, but also to denature elements of our self-understanding by placing them in the context of historical and social processes of the production of meaning. Self-creation and self-recreation, writing and re-writing of one's identity, are no longer seen as a privilege, but, at least theoretically, are within the reach of everybody. Identity seen as a work in progress blows up the stiff frames of socially and culturally constructed roles and makes room for self-reflexivity, fluidity and openness. Parkins and Craig analyze the phenomenon of slow living as a form of promotion of a "position counter to the dominant value-system of 'the times'"[65] and this allows them to move the analytical accent from discontents with global risk culture to a mobilization of practices which, although marked by an awareness of tensions and contradictions of high-modernity, are at the same time expressive of a growing valorization of "community, peace and slowness."[66]

The later chapters of *Slow Living* propose interdisciplinary readings of slow food, time and speed, space and place, food and pleasure and the politics of slow living, from the perspective of ideological embeddedness of the Slow Food and Slow Cities movements. Interestingly, Nick Osbaldiston, editor of *Culture of the Slow. Social Deceleration in an Accelerated World* (2013), extends his reading of slow living and claims that Parkins and Craig's analysis embraces too narrow a spectrum of cultural practices and is marked by too narrow a definition of slowness.[67] A polemical tone, which has recently invigorated debates about slowness, is struck by Tomlinson who is not willing to see slow living in terms of movements, while Osbaldiston is critical of Parkins and Craig's approach which he finds "unnecessarily dependent on analysing organised social movements."[68]

---

63 David Chaney, *Lifestyles* (London: Routledge, 1996), p. 52.
64 Parkins, Craig, *Slow Living*, p. 14.
65 Parkins, Craig, *Slow Living*, p. 1.
66 Parkins, Craig, *Slow Living*, p. 14.
67 Nick Osbaldiston, "Slow Culture: An Introduction," in *Culture of the Slow. Social Deceleration in an Accelerated World*, ed. Nick Osbaldiston (London: Palgrave Macmillan, 2013), p. 10.
68 Nick Osbaldiston, "Consuming Space Slowly," in *Culture of the Slow. Social Deceleration in an Accelerated World*, ed. Nick Osbaldiston (London: Palgrave Macmillan, 2013), p. 74.

He instead proposes to explore a wide range of activities which are expressive of individual engagement with present-day re-evaluations of the established benefits of accelerated lifestyles. According to him, "not all manifestations of the 'slow' need be founded in a dialectic of ethical and personal interest,"[69] and this claim signals a departure from Parkins and Craig's analytical model which he sees as excluding practices which are not motivated by ecologically or materialistically oriented values. What emerges as a central difference between Parkins and Craig's reading of slow living, and the perspective adopted by Osbaldiston, is that Parkins and Craig conceptualize those who are committed to the slow cause as self-reflexive subjects who "[b]y purposely adopting slowness, [...] seek to generate alternative practices of work and leisure, family and sociality"[70] and whose decision to engage with slow practices is motivated by moral and ethical reasons. Osbaldiston proposes a less rigorous definition of slowness and embraces a wide range of practices which include, among others, slow fashion, slow travel, cycling, lifestyle migration and sexual relations. The result is a collection of essays which approach cultural deceleration not so much in the context of the institutionalized and organized forms it can adopt but as a sweeping cultural response to accelerated lifestyles, and which extend the spectrum of practices to "include areas which at times appear apolitical."[71] This perspective puts Osbaldiston in a position to pose the question, absent from Parkins and Craig's analysis, whether slowing down is a revolution or "simply another 'phase' of consumer capitalism."[72]

Although theorists and cultural analysts of speed and slowness may differ in their understanding of the character of slow movements and activities and their power to reconfigure the already existing social structures and may want to include in their analyses different aspects of slow living, their conceptualizations of the phenomenon might help identify those aspects of it which may be recognized as defining. In representations popularized by practitioners of slow life and definitions proposed by its theorists, slow living appears as a set of social and cultural practices rooted in a desire to rearticulate one's relation with time by attentive and mindful engagement with the everyday. In an essay called "Out of Time: Fast Subjects and Slow Living,"[73] one of the first comprehensive theoretical

---

69   Osbaldiston, "Slow Culture: An Introduction," p. 11.
70   Wendy Parkins, "Out of Time: Fast Subjects and Slow Living," *Time & Society*, Vol. 13 No. 2/3 (2004), p. 364.
71   Osbaldiston, "Slow Culture: An Introduction," p. 10.
72   Osbaldiston, "Slow Culture: An Introduction," p. 11.
73   Parkins, "Out of Time: Fast Subjects and Slow Living," p. 364.

examinations of slow living, Wendy Parkins describes it as a form of "conscious negotiation of the different temporalities which make up our everyday lives, deriving from a commitment to occupy time more attentively."[74] The main implication of the above excerpt is expressive of the fact that none of the leading theorists and practitioners of slowness wishes to postulate a withdrawal from the world of speed and efficiency because "[s]low living is not a return to the past, the good old days (pre-McDonalds Arcadia), neither is it a form of laziness, nor a slow-motion version of life, nor possible only in romantic locations like Tuscany."[75] Parkins' description of slow living emphasizes a need to reprioritize the list of social and moral commitments and self-imposed obligations and make space for slowness in the hectic rhythm of life. In slow living the extremism of fast lifestyle is to be replaced by a sense of balance between the fast and the slow, which requires society's ability to negotiate between a need to accelerate and to slow down. Wolfgang Sachs suggests an implementation of what he calls a "politics of selective slowness"[76] based on the compromise between the fast and the slow which, although still guaranteeing delivery of fast service, may promote practices that permit sustainable development. "[S]low living is about installing the *possibility* [emphasis original] of slowing down,"[77] note Parkins and Craig, and their observation reflects a need to widen the range of lifestyle options that culture can accept.

Parkins' reading of deceleration, further developed in *Slow Living*, highlights the role which attention, often referred to as mindfulness, care and deliberation, plays in the reformulation of man's relation with time. What motivates and fuels slow practices and slow living is not so much a desire to have more free time, but rather to have more free time for meaningful activities. Parkins observes:

> Implicit in the practices of slow living is a particular conception of time in which 'having time' for something means investing it with significance through attention and deliberation. To live slowly in this sense, then, means engaging in 'mindful' rather than 'mindless' practices which make us consider the pleasure or at least the purpose of each task to which we give our time.[78]

---

74 Parkins, "Out of Time: Fast Subjects and Slow Living," p. 364.
75 Parkins, Craig, *Slow Living*, p. ix.
76 Wolfgang Sachs, "Speed Limits," in *Speed – Visions of an Accelerated Age*, eds. Jeremy Millar, Michiel Schwartz (London: The Photographers' Gallery and the Trustees of the Whitechapel Art Gallery, 1998), p. 132.
77 Parkins, Craig, *Slow Living*, p. 73.
78 Parkins, "Out of Time: Fast Subjects and Slow Living," p. 364.

Care and attention are represented as forming the foundation on which slowness may develop to become a "preferred mode for the heightened awareness."[79] In the above reflections slowness emerges as a crucial instrument in the process of liberation from the oppressiveness of fast time, which in the culture of instantaneity requires a state of permanent readiness to accomplish goals and be efficient. Parkins and Craig's reflections are permeated by a tone similar to the one which we find in a confession made by Sansot: "I […] would have liked to carve out a niche for myself, neither that of exile nor that of a retreat into an emptiness verging on non-existence or eternity, a place to shield me from being battered by a temporality *not of my making*" [emphasis mine].[80] He verbalized what has become a fundamental element of man's experience: a sense of loss of control over one's time. His craving for an escape from temporality *not of his own making* is expressive of a sense of helplessness and stress brought on by the fact that our time is controlled by an interconnected system of social structures, practices, organizations and institutions the efficiency of which is increased by homogenization of our experience of time. It is not the lack of time but rather a lack of control over it that has become a major source of anxiety for an increasing number of people. Stefan Klein builds a connection between a sense of living within the confines of timetables and deadlines, and a liberating sense of control over time when he says that "[…] we don't experience stress because time is in short supply, but rather because we think we have no control over our time."[81] Seen from this angle, slowing down may help reconstruct one's sense of agency and empowerment by offering a chance to reclaim one's control over time.

The phrase "slow is beautiful" reverberates across a multitude of diverse contexts and discourses, announcing a new episode in the history of the relationship with speed but as Sachs is careful to note, "[s]low, it turns out, is not just beautiful, but often also reasonable."[82] Sachs's claim connects with Shin'ichi Tsuji's understanding of the meaning of slow, for whom it also means "ecological" and "sustainable."[83] Analysing cycling as a manifestation of the culture of the slow, Martin Ryle and Kate Soper openly declare that going slow is not only a question of personal preferences and tastes but being now more than ever implicated in a

---

79   Parkins, Craig, *Slow Living*, p. ix.
80   Sansot, "On the Proper Use of Slownes," p. 299.
81   Stefan Klein, *The Secret Pulse of Time. Making Sense of Life's Scarcest Commodity*, trans. Shelley Frish (Frankfurt am Main: Da Capo Press, 2007), p. 202.
82   Sachs, "Speed Limits," p. 132.
83   Shin'ichi Tsuji, "Slow is Beautiful," p. 301.

complex system of interdependent structures, it should be understood as an obligation. They say:

> Environmental damage, above all carbon dioxide emissions, is the most evident cost paid for speedy transport. In the threats and discourses of climate change and 'peak oil,' contemporary affluent societies confront and half-acknowledge the fact that rather than speeding up, or even going on at today's pace, we are going to have to slow down.[84]

For Ryle and Soper, deceleration matters not only as an act of individual negotiation of the conditions of accelerated culture and a quest for alternatively defined pleasure but also as a set of cultural practices which seem to provide the only reasonable response to the prospects of ecological disaster we have been meticulously preparing for ourselves. The ecological orientation of many activities performed under the banner of slowing down, which they read in terms of culture's inevitable path, is embedded in the firm belief "that it is unlikely that a society which always moves in the fast lane can never be environmentally or even socially sustainable."[85]

The commitment of slow living to the cause of ecological and sustainable development, its orientation towards negotiation of irritations brought about by dominating models of consumer lifestyles, and a quest for the alternatively defined good life, places the issue of agency at the centre of the slow discourse. For Parkins and Craig slow living creates a space in which to "exercise agency over the pace of everyday existence and the movements across, and investments in, the respective domains of everyday life."[86] Agency in the context of slow living assumes the capacity to make a difference by making conscious choices: "[t]o declare the importance of slowness and to act upon it is a powerful expression of agency in a world that so strongly propels us in different directions."[87] A commitment to the "slow cause" is represented by Parkins and Craig as a manifestation of one's power to confront the accelerated rhythm of life and its discontents, and corresponds with Sansot's reading of "the proper use of slowness" when he says: "[…] the slowness of which I speak is not a character trait but a life choice."[88] Neither Parkins and Craig nor Sansot see slowness as "reducible to the

---

84 Martin Ryle, Kate Soper, "Alternative Hedonism: The World by Bicycle," in *Culture of the Slow. Social Deceleration in an Accelerated World*, ed. Nick Osbaldiston (London: Palgrave Macmillan, 2013), p. 97.
85 Sachs, "Speed Limits," p. 132.
86 Parkins, Craig, *Slow Living*, p. 67.
87 Parkins, Craig, *Slow Living*, p. 75.
88 Sansot, "On the Proper Use of Slownes," p. 298.

incapacity to sustain a faster pace."[89] Strategic slowness, that is to say, slowness implemented as a strategy of resistance, is not a result of the lack of speed but an expression of one's power to negotiate between various temporalities of global culture and a sense of agency which allows one to accept one's responsibility for the good life which slowness promises.

The issue of agency in the context of deceleration is, however, not unproblematic. The most significant aspect, which is often underlined by those who approach an alternative cultural offer proposed by the discourse of slow life with a dose of scepticism, concerns the kind of subjects who are able to get involved in slow practices. A prevailing criticism foregrounds the middle-class origin of practitioners of slowness and their financial security, which critics see as a condition of slow life, regardless of the proclaimed common availability of slow practices stressed in all representations of slow culture. To put it crudely, they see agency in the context of slow life as being available to those who can afford it and slowness itself as "just another 'fad' of the never-ending life projects of a middle class culture."[90] However, when one thinks about what Manuel Castells identified as one of the patterns in the development of human societies, which consists in a principle according to which "where there is domination, there is resistance to domination; where new forms of domination emerge, new forms of resistance ultimately surge to act upon the specific patterns of domination,"[91] the rise of the discourse of slowness and its popularity may be seen as part of a response to the domination of those models of life which have been most popularized among and embraced by affluent Euro-American middle classes. In view of Castells' interpretation of the dynamics of human development, the class character of slow living seems to be closely connected with the fact that the phenomenon of slow living is the product of those who are most influenced by the accelerated growth of consumer capitalism.[92]

Conclusions

Over half a century before the first publication of the Slow Food Manifesto, Harry Allen Ovestreet wrote that "[t]ime is no neutral thing that is what it is

---

89   Sansot, "On the Proper Use of Slownes," p. 298.
90   Nick Osbaldiston, "Conclusion: Departing Notes on the Slow Narrative," in *Culture of the Slow. Social Deceleration in an Accelerated World*, ed. Nick Osbaldiston (London: Palgrave Macmillan, 2013), p. 189.
91   Manuel Castells, *The Power of Identity* (Malden, MA.: Wiley-Blackwell, 2010), p. 147.
92   Osbaldiston, "Conclusion: Departing Notes on the Slow Narrative," p. 189.

despite what we are. Time, in large measure, takes on the complexion of ourselves. We can lengthen it or shorten it, fill it or empty it, make it into a thing of terror or into a rhythm of delight,"[93] and today the excerpt from *A Guide to Civilized Leisure* may be read as a declaration of faith by critics of modern time policy. The recent popularity of a new literature which valorizes slowness "in the midst of ever intensifying waves of texting, messaging, and tweeting"[94] is emblematic of a sense of disillusionment and weariness generated by the culture of speed. Interestingly, Honoré says that "[w]e have been here before,"[95] and Shin'tchi Tsuji admits that his declaration that "slow is beautiful" is just an act of "re-heating something old and familiar."[96] One may be tempted to approach the current interest in slowness in the context of a certain cultural and historical *déjà vu*, which consists in the return of a sense of anxiety generated by an acceleration experienced in all spheres of life. However, what shaped and defined the reactions of the opponents and critics of social and cultural acceleration at the beginning of the age of acceleration was primarily the fear of the unknown and an equally strong desire to maintain the *status quo*. Slowness was seldom celebrated for its inherent values, which could be recognized for their potential to suggest and inspire alternative lifestyles, subversion and contestation; low speed that was rather seen by critics of acceleration as a quality which would guarantee the stability of the existing order.

In the last few decades the term "slowness" has taken on wide valences in Western culture. A recent revalorization of slowness, this "slow revolution," as its enthusiasts put it, is a marker of a certain cultural solstice which post-industrial culture has been witnessing. A desire to liberate oneself from a sense of clutter generated by contemporary culture and the tyranny of fast time have introduced a number of practices which in certain social contexts can be seen as "a response to globalization and its perceived impact on everyday life."[97] Placing on their banners words of encouragement to slow down they create a space to articulate the need to follow a model of life different from the ones promoted by the culture of immediacy. The popularity of slogans which promote slowing down and the naturalness with which some of them have entered circulation in signifying

---

93   Harry Allen Overstreet, *A Guide to Civilized Leisure* (New York: Norton & Company, 1969), p. 164.
94   Jeffrey T. Schnapp, "Acknowledgements," in *Speed Limits*, ed. Jeffrey T. Schnapp (Milan: Skira, 2009), p. 8.
95   Honoré, *In Praise of Slowness*, p. 14.
96   Shin'ichi Tsuji, "Slow is Beautiful," p. 301.
97   Parkins, Craig, *Slow Living*, p. 2.

practices testify to their vitality and the emotional input which has the potential to produce a response by contemporary mis-men living in a globalized, accelerated reality. A recognition of the significance of deceleration by cultural theory illustrates on the other hand an understanding of the role slow time may play in the construction of social and cultural meanings and push forward a debate concerning the potential of slowing down to transform existing social structures and systems of value. The focus of slow living on pleasure chimes interestingly with the concept of alternative hedonism which conceptualizes pleasure as a means of social change, an aspect which will be the subject of discussion in Chapter Four.

Chapter Four: The Temporality of (Other) Pleasures

> And who knows if speed itself will become world weary?
> Valéry Larbaud, "Slowness"

> There are all sorts of pleasures that are actually going missing.
> Kate Soper, "An Interview"

Published in 1996, *Slowness* by Milan Kundera may be read as an introduction to the history of the relationship between speed and pleasure. The questions which the protagonist asks and the reflections they inspire illustrate the change in the ways in which the experience of pleasure has been reconceptualized over the course of time:

> Why has the pleasure of slowness disappeared? Ah, where have they gone, the amblers of yesteryear? Where have they gone, those loafing heroes of folk song, those vagabonds who roam from one mill to another and bed down under the stars? Have they vanished along with footpaths, with grasslands and clearing, with nature? There is a Czech proverb that describes their indolence by a metaphor: 'They are gazing at God's windows.' A person gazing at God's windows is not bored; he is happy. In our world, indolence has turned into having nothing to do, which is a completely different thing: a person with nothing to do is frustrated, bored, is constantly searching for the activity he lacks.[1]

The novel attempts to find answers to these questions by navigating between the present and the 18th century, the age of artificiality and fabrication, deliberately composed scenes and performed emotions, the age which achieved mastery in the art of remaining in a state of excitement and which possessed the wisdom of slowness. With the eighteenth-century culture situated in the background Kundera draws a picture of the present, which has offered itself to "a speed that is non-corporeal, non-material, pure speed, speed itself, ecstasy speed."[2] He perceives speed as being built up by a desire to forget: while memory reveals the ecstatic nature of time, the ecstasy of speed opens the door of forgetting. Cultural amnesia declares the past invalid and annuls responsibility for the future. Without access to the archives of memory the culture of speed forgets its own weariness; "caught in a fragment of time cut off from both the past and the

---

1   Milan Kundera, *Slowness*, trans. Linda Asher (London: Faber and Faber, 1996), pp. 4–5.
2   Kundera, *Slowness*, p. 4.

future,"[3] it is like a speeding motorcyclist for whom acceleration becomes the aim in itself.

The most recent decades have witnessed a return of the pleasures of slowness in the discourse of slow living. In the first part of this chapter I approach pleasure from the perspective of velocentrism in order to situate its history in the context of pleasure's relations with speed, both high and low. The emergence of new technologies of speed in modernism brought with it a new experience of pleasure and stabilized the connection between pleasure and speed, further enhanced and modified in consumer culture. Kate Soper's analysis of consumerism and the rise of alternative hedonism, which she explains in terms of dissatisfaction with consumer culture, supports an approach to deceleration from the perspective of a quest for an alternatively defined self-fulfilment which reconceptualizes the dominant definitions of the good life.

The focus of the second part of the chapter is on pleasures that have been redefined under an influence of slow living philosophy and on the ways they are represented in the discourse of slow and simple life. The pursuit of the pleasures of slowness, whose disappearance Kundera laments in his novel, emerges as a common denominator which unites not only those "who have opted for 'downshifting,' reduced working hours and more sustainable lifestyles,"[4] but also those who rediscover the pleasures of cycling or the slow contemplation of art. In the cases analyzed in this chapter a sense of (alternative) pleasure is determined by a conscious choice of one's model of relation with (slow) time and often helps create a space for resistance against the fast flow of contemporary life.

1. Between the Pleasures of Speed and Alternative Hedonism

The velocentric perspective allows one to read speed, both high and low, as a culture-forming principle which operates through a variety of fields, structures and practices. When pleasure is approached from the velocentric perspective three aspects emerge as central. Firstly, the perspective allows one to emphasize the constructed character of pleasure, which, as such a construct, has its own history and definitions which are culture and time specific. In other words, the perspective helps "historicise a subjective sensation,"[5] to borrow a phrase from

---

3   Kundera, *Slowness*, p. 3.
4   Kate Soper, "Introduction: The Mainstreaming of Counter-Consumerist Concern," in *The Politics and Pleasures of Consuming Differently*, eds. Kate Soper, Martin Ryle, Lyn Thomas (Basingstoke: Palgrave Macmillan, 2009), p. 6.
5   Enda Duffy, *The Speed Handbook. Velocity, Pleasure, Modernism* (Durham: Duke University Press, 2009), p. 4.

Enda Duffy. Secondly, a speed-oriented analysis of pleasure as a socio-cultural construct implies an evaluation of the extent to which speed has informed a sense of both pleasure and displeasure and affected their representations. Thirdly, the velocentric reading of pleasure presupposes a distinction between the pleasure of speed and the speed of pleasure, the significance of which consists in the fact that consumer culture has turned the speed of pleasure delivery into an indispensable element of consumer satisfaction and wellbeing. To read the processes of social and cultural construction of pleasure and its experience from the velocentric perspective is to accentuate their dynamic character and their responsiveness to changing cultural practices. The evolution of man's relation with speed and its accessibility have led to reformulations of the "mainstream" definitions of pleasure and to the relocation of their sources.

In her analysis of physical sensations as an aspect of mobility, and movement as the source of pleasure, Juliet Solomon refers to the pleasures of riding a roller coaster and a canal boat to claim that what these two forms of mobility have in common is that "the pleasure derived from them has its origins in the sensations (real or imagined) of physical movement."[6] Of these two "machines" the roller coaster is much more easily seen as a symbol of early Western prioritization of the pleasures of high speed. In a manner which correlates with Kundera's description of a motorcyclist immersed in the sensation of speed, Solomon writes:

> The roller coaster is about nothing except subjection to the sensation of movement. It does not matter where it is, and it does not matter where it goes. What matter are the changes in consciousness, which it triggers, as you are turned round about, shot into the air, thrust into the depths, suspended perhaps upside down, with your views inverted and your horizon blurred. For those who like it, it is enjoyable and, in a way addictive, the kind of experience which some might call a 'serotonin fix.'[7]

Although the roller coaster has always been thought to be a form of entertainment available to the masses, for centuries the pleasures of speed were elitist and reserved for those who could afford access to the means of fast transport, be they horse or coach. The roller coaster was perhaps the first example of modern technology which provided the public with an opportunity to enjoy the thrills of high speed as entertainment and which added an almost ecstatic immersion in speed to the widening spectrum of experiences of movement.

---

6   Juliet Solomon, "Happiness and the Consumption of Mobility," in *The Politics and Pleasures of Consuming Differently*, eds. Kate Soper, Martin Ryle, Lyn Thomas (Basingstoke: Palgrave Macmillan, 2009), p. 161.
7   Solomon, "Happiness and the Consumption of Mobility," p. 160.

What the roller coaster made available on a modest scale, the car put within the reach of almost everybody. It quickly found its place in the cultural landscape of the modern period by providing men and women with a way to increase their mobility on an unprecedented scale. It gave drivers a sense of freedom from trains and timetables and became an integral component of the definition of the good life, which construed wellbeing as including one's ability to move fast.[8] The car also created a desire to experience the new pleasures of speed, which stigmatized slowness as an obstacle to living in a technologically enhanced and a rapidly accelerating world. Delivered by new technologies of mobility, the new pleasures were accompanied by "defamiliarizing shocks" which were "directly physical rather than intellectual or aesthetic [...] and this made them immediately pleasurable, touching the body, potentially addictive."[9] Democratization of access to the "immediately pleasurable" experience of driving broadened the range of activities identified by Colin Campbell as traditional hedonism which is informed by a desire for pleasurable sensations. What "eating, drinking, sexual intercourse, socializing, singing, dancing and playing games"[10] have in common is that they cause pleasure through the provision of direct experiences. The somatic pleasure of driving was "in the first instance [...] *felt*" and as such "did not need to represent itself."[11] Duffy's study of the somatic aspects of the new pleasures read in the light of Campbell's analysis allows one to place the pleasures of driving within the framework of traditional hedonism and accentuate the role speed played in a new articulation and representation of the hedonist experience.

The claim made by Duffy that the common availability of speed in modernism marked a new stage in the history of impatience[12] affords one to conceptualize another turning point in the velocentric readings of pleasure and its representations, and situate it in the context of commodity culture which has reorganized the relation between speed and pleasure by adding a new dimension to it. To study the history of pleasure from the perspective of the ways its structures have been related to speed means to recognize the significance of the fact that in consumer culture the pleasures of speed run in tandem with the immediacy of pleasure delivery. The rise of modern hedonism has been marked by a departure

---

8   John Urry, *Mobilities* (Cambridge: Polity Press, 2007), p. 117.
9   Duffy, *The Speed Handbook*, p. 5.
10   Colin Campbell, *The Romantic Ethic and the Spirit of Modern Consumerism* (Oxford: Blackwell, 1987), p. 69.
11   Duffy, *The Speed Handbook*, p. 4.
12   Duffy, *The Speed Handbook*, p. 66.

from the immediacy of sensory experience and replaced by a drive towards a construction of mental images consumed "for the intrinsic pleasure they provide."[13] A modern hedonist "withdraw[s] from reality as far as he encounters it, ever-casting his day-dreams forward in time, attaching them to the object of desire and then subsequently 'unhooking' them from these objects as and when they are attained and experienced."[14] Viewed from this angle,[15] the modern consumer is suspended between action and inaction: while withdrawn "from reality," he can exercise a creative potential to invest the props of consumer culture with new meanings, therefore becoming an "artist of the imagination, someone who takes images from memory, or the existing environment, and rearranges or otherwise improves them in his mind in such a way that they become distinctly pleasing."[16] Modern hedonism is thus marked by a quest for goods and services which, in the words of Roberta Sassatelli, are "ripe for personal creative fantasy."[17]

At the intersection of modern hedonism, consumer creativity and the dominance of the principle of immediacy, the bond between speed and pleasure has grown even stronger because, as Martin Ryle and Kate Soper observe, "[c]onsumer culture glorifies speed."[18] In an accelerated culture the sensation

---

13   Campbell, *The Romantic Ethic and the Spirit of Modern Consumerism*, p. 77.
14   Campbell, *The Romantic Ethic and the Spirit of Modern Consumerism*, pp. 86–87.
15   While Campbell's analysis helped identify the main concerns of consumption studies and articulate their main conceptual perspectives, recent critical reflections on the nature of consumer practices have been marked by a growing number of "revisionist" interpretations of consumption which see it as the site of an active engagement with the everyday and a space where civil consciousness may manifest itself. Polemical debates about the active/passive nature of consumer practices intensified after September 11th: calls made to encourage shopping as a way to manifest American invincibility gave Americans a sense of empowerment they needed so much. On the other hand, Naomi Klein claims that "as more people discover the brand-name secrets of the global logo-web, their outrage will fuel the next big political movement, a vast wave of opposition squarely targeting trans-national corporations, particularly those with very high name brand recognition." Naomi Klein, *No Logo* (London: Harper Collins, 2000), p. xviii. In the light of the above, contemporary consumer practices can be seen as a form of expression of one's political and social views and a form of active participation in the politics of the everyday.
16   Campbell, *The Romantic Ethic and the Spirit of Modern Consumerism*, p. 78.
17   Roberta Sassatelli, "Consumer Identities," in *Routledge Handbook of Identity Studies*, ed. Anthony Elliot (New York: Routledge, 2011), p. 237.
18   Martin Ryle, Kate Soper, "Alternative Hedonism: The World by Bicycle," in *Culture of the Slow. Social Deceleration in an Accelerated World*, ed. Nick Osbaldiston (London: Palgrave Macmillan, 2013), p. 95.

of pleasure relies heavily on the quickness with which it is provided; to put it differently, the speed at which pleasure may be achieved has become an integral element of the way pleasure is conceptualized. For Tomlinson "[t]he attractions of immediacy are strong because they *do* [emphasis original] deliver the comforts, conveniences and pleasures that have been long promised in the cultural imagination of modernity."[19] The pleasures of speed as a sensory experience have paved the way for the pleasures of immediacy which today embrace both sensory experiences and narratives of individual dreams and imagination.

Strong and attractive as they might be, the appeals of immediacy have been recently at the centre of ongoing debates and reflections about life in high modernity. We have been witnessing a growing interest in practices and narratives of contestation that undermine what consumer ideology represents as the ideal of the good life. They are beginning to leave a mark on the character and tone of debates about consumption and consumer lifestyles in the humanities, cultural studies and sociology which highlights the need to redefine and reconceptualize self-fulfilment and life-satisfaction in the world of pleasures, "each more seductive than others and each pushing its predecessor into early retirement and ultimate oblivion."[20] A space has opened up for the critical re-readings of consumer ideologies: the experience of man has been increasingly marked by a sense of a tension between the ideals of happiness and fulfilment, endlessly promised by cultural, political and financial institutions, and the realities of everyday life, which often brings a sense of being oppressed by a dystopian dream which has come true – a dream of unlimited choice of goods. The quality of everyday life has thus turned out to be a way to verify an ideal whose premise is that unlimited consumption, technological development and acceleration guarantee both individual satisfaction and global welfare. However, it is telling that the tension and the anxieties it has released are no longer represented as an unavoidable price which has to be paid for the comforts of life, which have been repeatedly defined as availability of goods and immediate gratification of needs.

The growing anxiety of those who feel disillusioned by an accelerated consumerist lifestyle has recently become visible in various aspects of life. The members of a growing "trans-class class of the 'disillusioned seduced,'"[21]

---

19   Tomlinson, *The Culture of Speed*, p. 158.
20   Zygmunt Bauman, *44 Letters from the Liquid Modern World* (Cambridge: Polity Press, 2010), p. 47.
21   Kate Soper, "Alternative Hedonism, Cultural Theory and the Role of Aesthetic Revisioning," *Cultural Studies*, Vol. 22, No. 5 (2008), p. 573.

aware that the costs of the luxuries and enjoyments of living in an accelerated and technologically advanced reality may be high, have started to demand a platform where their concerns and anxieties can be articulated. In stark contrast to the promises of the ideologies of neo-liberal economies, it seems certain that immediacy, prosperity and abundance have failed to guarantee unproblematic and tension-free existence. Relatively easy access to activities and routines which not so long ago were available only to an affluent elite has recently produced a disturbing sense that the pleasures they offer are often "compromised by their negative effects" and that they are "pre-empting other enjoyments."[22] Below the surface of affluence and immediate availability of goods and services there lurks "an intuition of *the other pleasures* [emphasis mine] that it [a consumerist lifestyle] constraints or destroys, especially those that would follow from a slower, less work-dominated pace of life."[23] Such an intuition testifies to a desire to revalorize those pleasures which have been stigmatized as un-modern and unaffordable in dominating ideology of the "time-is-money" culture. "What-you-want-and-when-you-want-it"[24] types of pleasures, which have monopolized the imagination of pleasure-seekers, provoke, rather unexpectedly for a self-confident Westerner, a sense of discontent if they reveal their socially and environmentally negative consequences. Soper's analysis of the present cultural agitation and ferment simmering under the surface of affluence and comfort leads her to articulate the concept of alternative hedonism as an expression of the quest for alternative points of reference in the world in which "traditional" frameworks can no longer be read as symbols of stability and security. Her study of consumer and anti-consumer practices has inspired her to place an analytical focus on the culture-forming potential of alternative pleasures and on their role of a trigger of social transformation. The fact that Soper links peoples' desire to return to pleasures long forgotten by the culture of excessive production and consumption with the possibility of a wider social change is important as it allows us to examine cultural aspects of deceleration within the framework of alternative hedonism and its rhetoric.

The pursuit of alternatively defined pleasures illustrates a sense of disillusionment with the promises of consumer ideologies, according to which the consumption of goods and services is a way to enhance a sense of happiness and life satisfaction. A desire to "live differently" is framed by the sense of loss

---

22   Soper, "Introduction: Mainstreaming…," p. 4.
23   Soper, "Introduction: Mainstreaming…," p. 4.
24   See: Zygmunt Bauman on virtual sex in *44 Letters from the Liquid Modern World* (Cambridge: Polity Press, 2010), pp. 22–26.

experienced by a growing number of consumers who wish to "advance beyond a mode of life that is not just unsustainable but also in many respects unpleasurable and self-denying."[25] Alternative hedonism redefines pleasure by pointing to the possibility that it can be liberated from the confining and one-dimensional rhetoric of consumerism. It embraces the decolonization of desire from the autocratic rule of consumer philosophy of self-fulfiment and satisfaction attainable along the lines of ready-made consumer recipes for individual and communal wellbeing. Central to Soper's reconceptualization of pleasure is the stress she puts on the issue of self-interest, which she identifies as a motive factor of any social shift. She is convinced that unless consumers are presented with an option which would appeal to them and provide a sense of moral, financial and psychological satisfaction, chances for the change and a departure from dominating models of consumer happiness are minimal. Soper proposes the creation of a "seductive alternative conception of what it is to flourish and to enjoy a high standard of living."[26] Alternative hedonism is underpinned by a conviction that an escape from the consumerist model of the good life, whose discontents and displeasures have become conspicuous is not only possible but also necessary.

"Consuming differently," Soper claims, may create a "new erotics of consumption or hedonist 'imaginary.'"[27] Her study of alternative hedonism registers the rise of the new rhetoric which is concomitant with a change of the focus which the will to live differently brings: from a sense of threat and exhaustion to a sense of celebration, fulfilment and pleasure. Soper believes that "[a]lternative hedonists can speak more compellingly, and persuasively, than the prophets of environmental catastrophe."[28] The concept of sustainable hedonism, proposed by Marius de Geus, evidences this rhetorical change when, in a somehow utopian manner, he predicts that

> [i]n a post-consumerist society the main emphasis will be not on the outward manifestations of status and success, but on cultivating the inward aspects of human wellbeing. One may cite here the pleasures of relaxation and balance, of attending more closely to our fellow creatures (both human and non-human), the enjoyment of meaningful work, of contributing to the community and of a general spiritual wellbeing [...].[29]

---

25 Soper, "Introduction: The Mainstreaming…," p. 3.
26 Soper, "Introduction: The Mainstreaming…," pp. 3–4.
27 Soper, "Alternative Hedonism, Cultural Theory…," p. 571.
28 Soper, "Introduction: The Mainstreaming…," p. 3.
29 Marius de Geus, "Sustainable Hedonism: The Pleasures of Living within Environmental Limits," in *The Politics and Pleasures of Consuming Differently*, eds. Kate Soper, Martin Ryle, Lyn Thomas (Basingstoke: Palgrave Macmillan, 2009), p. 121.

The change of the rhetoric as a result of a meaningful shift from apocalyptic visions to images of pleasure and a sustainable future corresponds to the shift from a sense of helplessness in the face of pending social and economic exhaustion to a sense of empowerment which the new rhetoric and imaginary both articulate and emphasize, since alternative hedonism "points to new forms of desire, rather than fears of ecological disaster, as the most motivating force in any shift towards a more sustainable economic order."[30] When Soper says that she "want[s] a political imaginary that highlights the sensuality, the almost baroque pleasures that we might otherwise indulge in,"[31] she does not intend to promote a withdrawal from the anxieties of the present and a retreat into hedonist amnesia. On the contrary, the stress on "cultivating," "attending" and "contributing to the community" relocates the sense of pleasure from the level of the individualistic particularity of one's own concerns and interests to the sense of communal wellbeing which becomes a powerful motive factor generating forms of active engagement with everyday life. This new alternative is seductive not only in terms of an individual sense of satisfaction brought about by a rediscovery of pleasures access to which has been made difficult in recent decades because of rapid technological growth, but also in the contexts of "altruistic compassion and environmental concern."[32]

Alternative hedonism puts forward a new discourse of wellbeing in which a central position is reserved for "more self-regarding gratifications of consuming differently"[33] which embrace, among other things, the pleasures of alternative temporalities, because, as Soper observes

> [i]f affluent societies are successfully to meet the environmental and social challenges of the future, they need to begin now to plan for a shift to a more readily reproductive way of living, to a low- or no-growth economic model rooted in an expansion of leisure time and rather different conceptions of social flourishing and human wellbeing.[34]

The phenomenon of slow living, with its reconceptualizations of the good life and strategies of resistance against the discontents of globalized and accelerated development which it promotes, can be viewed as embracing the postulates of

---

30   Soper, "Introduction: The Mainstreaming…," p. 3.
31   Kate Soper, "Interview. An Alternative Hedonism," interviewed by Ted Benton, *Radical Philosophy*, 92 (1998), p. 33.
32   Soper, "Introduction: Mainstreaming…," p. 4.
33   Soper, "Introduction: Mainstreaming…," p. 4.
34   Kate Soper, "Humanities Can Promote Alternative 'Good Life,'" *The Guardian*, November 30, 2010, accessed 17 April 2014, http://www.theguardian.com/commentisfree/2010/nov/30/humanities-promote-alternative-good-life

alternative hedonism and redefining social and cultural aspects of pleasure. In rediscovering the pleasures of slowness, Western hedonism may be seen as entering a new phase.

2. Slow Living and New Representations of Pleasure

While modernity construed the new pleasure as being informed by speed, high modernity, although still a site of processes which reinforce tight bonds between speed and pleasure, has witnessed a rise of practices which have been motivated by the wish to foreground the pleasure of slowness and have invested this concept with a variety of values and meanings. The issue of the relation between speed and pleasure returns, brought into the spotlight of critical attention and popular practice by the phenomenon of slow life. The rhetorical shift behind the claim that "slow is beautiful," which signals a departure from "traditional" understanding of slowness as a manifestation of inefficiency and backwardness, and the popularity of new conceptualizations of the good life, represent a new consciousness which undermines the assumptions of consumerist lifestyle and promotes alternative concepts of pleasure and self-fulfilment. Read from the perspective of alternative hedonism, slow living situates itself in the space whose contours are marked by a quest for pleasures "destroyed, pre-empted and threatened by hegemonic forms of development."[35]

The pleasure principle formulated by Soper allows her to look for the manifestations of alternative hedonism in "the expressed interests in the less tangible goods such as more free time, less stress, more personal contacts, a slower pace of life, etc., as lending support to criticism of the narrow materialism of consumer culture."[36] One of the premises on which a conception of the alternative good life has rested is that the understanding of growth and wellbeing needs to be disentangled from its immediate association with concepts which are often used as their synonyms, that is, acceleration and time management. Martin Ryle and Kate Soper observe that

> [t]he bid to do things faster, and thus reduce time spent on any given activity, is at the heart of the consumerist dynamic, whether it be a matter of information technology or of physical transport. The ideas of 'progress' and 'development' have become more or less synonymous with those of saving time or speeding up [...]. The tacit assumption in this association of human advancement with increased speed is that the faster we travel and communicate, the more exciting life will become, and the fuller and richer our experience will be.[37]

---

35   Ryle, Soper, "Alternative Hedonism: The World by Bicycle," p. 94.
36   Soper, "Alternative Hedonism, Cultural Theory...," p. 573.
37   Ryle, Soper, "Alternative Hedonism: The World by Bicycle," p. 95.

Slow living challenges such representations of the good life by offering a reconnection with pleasures which have been removed from the spectrum of contemporary experience. What characterizes the way in which the pleasures of slowness are represented in the discourse of slow living is their common availability. This attribute is emphasized by the practitioners of slow life whenever they want to challenge the accusation that the movement has an elitist character, which is often put forward by those who see slow living as a product of the affluent middle classes. In the words of Silvija Davidson, chair of Slow Food UK, "the right to pleasure is a leveller because it's shared by everyone. The point about slow is that it doesn't necessarily require radical change. By definition, it's gentle. You can do a little bit at whatever pace you want."[38]

Wendy Parkins and Geoffrey Craig's claim that slow living is an "attempt to live in the present in a meaningful, sustainable, thoughtful *and pleasurable* [emphasis original] way"[39] chimes with the premises of alternative hedonism and its recognition of the power of pleasure to redefine social structures and practices. Such a mode of thinking about pleasure and the role it can play in challenging the oppressive character of an accelerated globalized world was first articulated, explicitly (and with an emotional power which immediately triggered equally emotional response) in the Slow Food Manifesto. While from the perspective of the mis-man deceleration is always anti-progressive and little more than a self-banishment to the margins of the fast world and withdrawal from the (fast) game, Carlo Petrini's project rehabilitates slowness which the members of the Slow Food movement proposed as a crucial tool in the battle for a new social order. The manifesto is in fact a powerful declaration of a reflexive concern about the quality of future lives and of future-oriented thinking. The philosophy of the Slow Food movement moves beyond the particularity of individual preferences and tastes and while stressing the power of subjective pleasure and its potential for resistance, it suggests a possible course of change in the modes of production and consumption. What is practiced under the banner of deceleration is often motivated by the conviction that we need "not to sustain and hand on to future generations a living standard as currently defined, but to consume differently now in order to accommodate the goods [...] that are currently being lost or marginalized by 'high' standards of living."[40]

---

38   Silvija Davidson, quoted by Mark Palmer in "How to Go Slow," *The Telegraph*, July 26, 2008, accessed 28 March 2014, http://www.telegraph.co.uk/news/features/3637476/How-to-go-slow.html
39   Wendy Parkins, Geoffrey Craig, *Slow Living* (Oxford: Berg, 2006), p. ix.
40   Kate Soper, "Rethinking the 'Good Life': The Citizenship Dimension of Consumer Disaffection with Consumerism." *Journal of Consumer Culture*, Vol. 7(2), (2007), p. 211.

The manifesto starts with a diagnosis of the conditions of contemporary culture and lifestyles: "[w]e are enslaved by speed and have all succumbed to the same insidious virus: *Fast Life*, which disrupts our habits, pervades the privacy of our homes and forces us to eat Fast Foods."[41] The rhetoric of oppression, suspended upon the images of slaves and viruses, makes it clear that decisive action is needed in order to rescue endangered humankind, whose integrity is threatened by speed. The apocalyptic language used in the manifesto exudes a sense of exhaustion and disillusionment which is experienced by a growing number of people in affluent Euro-American culture. The almost dystopian imagery that is successfully employed by the rhetoric of the movement tells the story of a dream turned into a nightmare. Soper's analysis of the specificity of a present cultural moment reveals a similar concern when she says that there are the "signs that the affluent lifestyle is generating its own specific forms of disaffection."[42] The apparently smooth surface of high-standard living begins to display cracks and openings behind which lurks a sense of the end of a utopian desire for the good life that is presented by an accelerated world of consumption and globalization.

The solution, according to Petrini, consists in cultural deceleration, which is conceptualized as a point of departure for the implementation of further changes: "[t]o be worthy of the name, *Homo Sapiens* should rid himself of speed before it reduces him to a species in danger of extinction" (*SFM*, xxiii), we read in the manifesto. On the side of the slow life, whose contours have been sketched by the founding fathers of the Slow Food movement, there is a promised chance for a better future, which, however, requires a thorough redefinition of the concept of the good life. Essential to this conceptual change is recognition of the autocratic rule of speed and then readiness to employ slowness as a strategy of resistance against the artificially fast and still accelerating pace of life, a strategy which may restore the value of pleasure in the life of an individual:

> A firm defence of quiet material pleasure is the only way to oppose the universal folly of Fast Life. May suitable doses of guaranteed sensual pleasure and slow, long-lasting enjoyment preserve us from the contagion of the multitude who mistake frenzy for efficiency (*SFM*, xxiii).

Petrini's original emphasis on pleasure, put forward in the Slow Food Manifesto, was further developed and theorized in the "Preface" to *Slow Food. The Case for*

---

41 "The Official Slow Food Manifesto," in Carlo Petrini, *Slow Food. The Case for Taste*, trans. William McCuaig (New York: Columbia University Press, 2003), p. xxiii. Throughout the chapter the Slow Food manifesto will be referred to as *SFM*.

42 Soper, "Alternative Hedonism, Cultural Theory…," p. 571.

Taste where he examined the power of pleasure to initiate social change. His valorization of pleasure was not motivated by "pure hedonistic enjoyment for its own sake," but by recognition of the fact that a connection of pleasure to an "awareness and responsibility, study and knowledge"[43] may bring about a redefinition of existing social structures of food production and consumption. The moment of transformation could start when a sense of dissatisfaction with existing consumer pleasures begins to provide a foundation for their critical re-examination. The consumer lifestyle, with its stress on the instant gratification of needs and easily accessible pleasures, reveals internal tensions which point at a disturbing disaffection that more and more often casts a shadow on the present-day vision of an accelerated and globalized paradise.

Petrini's most comprehensive interpretation of the central role of pleasure in his project is to be found in the chapter called "Pleasure Denied, Pleasure Rediscovered." It starts with a quote from an essay "Fisiologia del piacere" ("The Physiology of Pleasure") by Giorgio Bert, who writes that "the word 'pleasure' still has a faintly dubious ring."[44] A commitment to pleasure, Bert noted, is read as a social disqualifier, a sign of one's inability to engage in most desirable processes of corporate competitiveness and work efficiency. Refraining from pleasure is evidence of discipline and a sense of responsibility because we have been taught to treat pleasure "as somehow artificial, a luxury for a few – most likely undisciplined and shameless – people."[45] Petrini airs a similar opinion when he says that "pleasure was, and is, a thorny subject" and that people "regard an interest in pleasure as a sign of superficiality."[46] One may be tempted to think that, probably for the sake of the power of rhetorical expressiveness, both Bert and Petrini deliberately dismiss the ambiguity of the status of pleasure in consumer culture. While it is true to say that to confess one's dedication to pleasure may threaten one's social status, we need to remember that as consumers we *are* pleasure-seekers and this aspect has been dismissed from the valorization of pleasure that Petrini articulates. It is not abstention from pleasure but rather the redefinition of types of pleasures available to consumers that defines contemporary life. A sense of pleasure has not been removed from the spectrum of modern experience, testimony to which is provided by both traditional and modern hedonism. Rather, it would be more appropriate to talk about

---

43   Carlo Petrini, *Slow Food. The Case for Taste*, trans. William McCuaig (New York: Columbia University Press, 2003), p. xvii.
44   Petrini, *Slow Food: The Case for Taste*, p. 20.
45   Giorgio Bert, quoted by Petrini in *Slow Food: The Case for Taste*, p. 20.
46   Petrini, *Slow Food: The Case for Taste*, p. 20.

prioritization of certain pleasures at the cost of marginalization, or repression, of others.

The focus Petrini places on the role of pleasure in the process of individual and social change sheds new light on the political power of pleasure to cause wider social change. It also underscores a correspondence between Petrini's interpretation of the consumerist lifestyle and Soper's alternative hedonism and its stress on "the sensual and spiritual pleasures of escaping the dominant model of the good life and calling for a cultural revolution in our perceptions of prosperity as the necessary first stage in building a mandate for a fairer and more sustainable economic order."[47] In Petrini's project sensual pleasure, especially that delivered by the sense of taste, may become the activator of social transformation. Petrini famously announced that "Slow Food endorses the primacy of sensory experience and treats eyesight, hearing, smell, touch, and taste as so many instruments of discernment, self-defence and pleasure. The education of taste is the Slow way to resist McDonaldisation."[48] The devaluation of man's sensory skills is seen by Petrini as the price that has been paid for the globalized production and consumption that have been recognized as markers of progress and tokens of wellbeing. Petrini posits a connection between the accelerated tempo of life and the developmental regress of our senses which is accompanied by a homogenization of taste. Moreover, his interpretation of the role of senses seems to challenge the Cartesian mind-body duality in a manner that corresponds with Ludwig Feuerbach's defence of the role the sensual plays in the process of thinking. In *Lectures on the Essence of Religion* (Lecture 10) he wrote:

> I regard the sensuous world as the first element of psychology and of philosophy in general; […] I cannot derive my body from my mind – for I have to eat or to be able to eat before I can think; as the animals demonstrate, I can eat without thinking, but I cannot think without eating; I cannot derive my senses from my faculty of thought, from my reason – for the reason presupposes the senses, but the senses do not presuppose reason […].[49]

Feuerbach's insistence on the philosophical dimension of food recasts the Cartesian apothegm. As Anthony Synnott writes: "for Feuerbach, the first principle is not '*Cogito*', but '*Senteo, ergo sum*': I feel, therefore I am."[50] In light of

---

47 Soper, "Humanities…,"
48 Petrini, *Slow Food: The Case for Taste*, p. 69.
49 Ludwig Feuerbach, quoted by Anthony Synnott in *The Body Social. Symbolism, Self and Society* (London: Routledge, 1993), p. 142.
50 Synnott, *The Body Social*, p. 143.

the above, Petrini's ideological framework seems to derive from Feuerbach's restoration of the value of the senses; in the discourse of slow culture it is *"Gusto ergo sum,"* "I taste therefore I am," to refer to a phrase popular among food experts, food philosophers and thousands of food lovers, which illustrates a turn towards rehabilitation of the sense of taste.

For Petrini the pursuit of pleasure has both an individual and a collective dimension. Rooted deeply in a sense of dissatisfaction with the taste of food produced by global companies, the new politics of taste transgresses the boundaries of individual space and becomes a manifestation of concern with the quality of life of the individual and the community as well as of concern with the condition of the natural environment and local areas. Pleasure redefined along the lines of individual and collective welfare becomes an expression of deep engagement and reveals a new dimension of alternative hedonism, turning it into a tool of social and cultural change. Petrini and Soper share the conviction that a sense of pleasure – redefined pleasure – can lead to liberation from the oppressive confines of a lifestyle imposed by an accelerated globalized culture and, to quote Parkins and Craig, "restore agency to subjects inundated by the sensory profusion of late capitalism in which pleasure may be all too fleeting."[51] "A new erotics of consumption" suggests the exhaustion of traditional modes of satisfaction on one hand, and on the other exposes the seductive appeal of alternative hedonism. In the discourse of slow living the private and the personal emerge as a site of political confrontation, contestation and negotiation. Petrini and Soper recognize the power of "self-interested concerns"[52] to initiate new cultural practices and trigger social changes, the desire for which is often rooted in dissatisfaction with those aspects of the present that are experienced as oppressive and alienating.

A redefined sense of pleasure is brought to the foreground of the velocentric analysis not only by slow living but also by the discourse of downshifting, which represents simplicity as a "part of getting off the fast track"[53] while creating a space to confront the requirements of consumer culture and challenge commodity fetishism and the tyranny of fast time by adopting a "less is more" philosophy. Petrini's call to "oppose the universal folly of Fast Life" and defend "quiet material pleasure" (*SFM*, xxiii) repeats some of the main assumptions put forward by downshifters and minimalists. They mark the contours of a phenomenon whose

---

51 Wendy Parkins, Geoffrey Craig, *Slow Living* (Oxford: Berg, 2006), p. 97.
52 Soper, "Alternative Hedonism…," pp. 571–572.
53 Elaine St. James, *Simplify Your Life. 100 Ways to Slow Down and Enjoy the Things that Really Matter* (New York: Hyperion, 1994), p. 60. Throughout the chapter this book will be referred to as *SYL*.

concerns have been articulated with such determination and urgency that it has been recognized by cultural theorists, sociologists and philosophers as an important form of response to the accelerated and excessive character of consumption. The philosophy of simple living takes various forms and ranges from downshifting or voluntary simplicity to minimalism. What all these practices have in common is that in an age of excessive consumption, production and work overload they derive their strengths and dynamics from the conviction that human satisfaction and a sense of wellbeing do not have to be determined by the accumulation of material goods and a financial preparedness to engage in all forms of pleasure, the range of which consumer culture has been constantly extending.

Simplify Your Life. 100 Ways to Slow Down and Enjoy the Things that Really Matter by Elaine St. James is one of many popular self-help books which promote the ideas of simple living and offer practical tips to be used in everyday practice of simplicity. The analysis of representations of an alternative sense of pleasure which the book articulates helps identify what Soper perceives as "motives for changing consumption practice that derive from the more negative aspects for consumers themselves of their high-speed, work-dominated, materialistic lifestyle," and which are "fed by a sense that important pleasures and sources of gratification are being lost or unrealised as a consequence of it."[54]

St. James is similarly motivated when she says that "we've have been compelled in recent years to go, to do, to be on the move, to experience all that the money can buy. Oftentimes, in the process, the things we really like to do have been overlooked" (*SYL*, 60). Her meeting with a group of professionals resulted in their composing a list of activities in which they would like to engage if only they had time to do so. The "dream list" included "watching the sunrise, walking on the beach, spending quiet time with the family or browsing in a bookstore," as well as "[s]itting quietly in a favourite chair and doing nothing" (*SYL*, 61). What the activities listed by St. James had in common was that they were sources of "the simple pleasures" (*SYL*, 61), access to which brings a reconceptualization of luxury which for downshifters no longer denotes exclusive and expensive purchases and leisure or entertainment. Those who consciously decide to slow down and lower their financial status reconstruct the hierarchies of values by replacing the imperative of keeping up with the Joneses with an openly expressed desire to "break out of that loop."[55]

---

54 Soper, "Re-thinking the 'Good Life'...," p. 211.
55 Carey Goldberg, "Choosing the Joys of a Simplified Life," *The New York Times*, 21 September 1995.

The essence of downshifting does not consist in a rejection of social, professional, communal or family obligations but rather in "doing them differently,"[56] St. James puts it in *Simplicity. Easy Ways to Simplify and Enrich Your Life*, another self-help book teaching the arts of simple living, and it is precisely in the concept of difference that the philosophy of simplification as an alternative lifestyle finds its ideological framework. "Doing things differently" implies the ability and readiness to negotiate socially constructed and commonly accepted codes and commandments and shape them so that they do not confine individual human potential but rather amplify it and create conditions favourable for its growth. What is vital for St. James is the ability to (re)construct one's own agency through developing a critical approach towards the dictates of popular lifestyles. She is aware that "going against the tide" (*SEW*, 51) requires courage and determination but she is also aware that it builds up the sense of "the freedom that comes from actively creating own interpretation of success" (*SEW*, 51–52). The downshifters' confrontation with the tyranny of commercially determined ideals exemplified by, among other things, the "idea that our clothes have to be whiter than white, that our blues have to be bluer than blue, that our mirrors have to shine with a brilliant lustre, or that our hardwood tabletops have to be polished to a blinding brightness" (*SEW*, 12) leads to a redefinition of the significance of social expectations and allows one to reinterpret the meaning of success and set one's own standards of wellbeing.

Another crucial aspect of the decision to simplify has been a sense of exhaustion caused by the permanent state of vigilance induced by the consumerist imperative to take advantage of all that the world of material culture and consumer lifestyle has to offer. St. James writes: "[n]ever before in the history of mankind have so many people been able to have so much, go so many places, and do so many things. We've worn ourselves out trying to have it all" (*SEW*, xviii). Living in a world of infinite possibilities, where everything informs the individual that an opportunity not used is an opportunity missed and wasted, generates not only material but also emotional and psychological clutter. "There are millions of other people out there who believe, as I did, that we can do it all, have it all, be it all" (*SEW*, 32), says St. James. The sense of social security, a vital element of which depends on being accepted by other members of one's community and the readiness to conform to commonly accepted standards, is defined now, more than ever before, by material culture. Modern consumers are kept in

---

56   Elaine St. James, *Simplicity. Easy Ways to Simplify and Enrich Your Life* (London: Thorsons, 1997), p. 12. Throughout the chapter this book will be referred to as *SEW*.

a state of permanent hunger for more and it is this insatiable desire that keeps the consumer market in motion. The consumers' path seems to have no horizon.

Downshifters' re-examination of their needs is indicative of a desire to "step off the consumer treadmill" (*SEW*, 81) and to slow down because the pace of life is recognized by them as a main obstacle on the way to achieving and enjoying the benefits of a simple life. Significantly, what the practitioners of voluntary simplicity often stress is this life's engagement not only with the material aspects of the everyday but also with inner, spiritual transformation and personal growth. While *Simplify Your Life* and *Simplicity* offer practical ways to simplify one's life in everyday activities and routines, *Inner Simplicity*, as the subtitle informs us, presents "100 ways to regain peace and nourish your soul."[57] *Inner Simplicity* is an illustration of the quest for spirituality and inner balance in a post-consumerist environment, represented in a manner that brings to mind de Geus' vision of a future based on cultivation of the "inward aspects of human wellbeing." The rhetoric which the discourse of simple living uses is in stark contrast with the disturbing and unsettling rhetoric of cultural debates about the present conditions of humanity which revolve around the sense of exhaustion, instability, crisis and fragmentation. The rhetoric of slow and simple living constructs an imagery which takes its strength and vitality from an accentuation of personal agency and a sense of empowerment as well as from the possibility to build a system of personal defence against what the advocates of both movements see as the alienating and destabilizing forces of the culture of speed and thus reclaim pleasures thought to have been lost.

The rhetoric which underlies most popular and accessible versions of voluntary simplicity, like the ones popularized by St. James, is informed by almost religious overtones which stress "get[ting] in touch with your intuition" (*IS*, 194), "learn[ing] to listen to your own voice" (*IS*, 208) or starting "inner exploration" (*IS*, 225), and may suggest that what simple living offers is a *simplified* version of inner spirituality. Interestingly, although not unexpectedly, the spirituality that the discourse of simple living proposes is represented as a form of inner growth which may be achieved *fast* and in a somehow *effortless* manner. The one hundred ways to achieve the state of inner peace, which the book promises, are put forward in the form of one hundred short mini-chapters, or extended manifestoes, which represent inner simplicity as something within the reach of everybody,

---

[57] Elaine St. James, *Inner Simplicity. 100 Ways to Regain Peace and Nourish Your Soul* (New York: Hyperion, 1995). Throughout the chapter this book will be referred to as *IS*.

as long as he follows the tips contained in the chapter titles such as "Slow Down," "Connect with the sun," "Smile a lot," or "Create joy in your life." Difficult as it may be, if we succeeed in putting aside self-imposed associations with often-ridiculed rhetoric and imagery of various New Age movements, we may note the ways in which the discourse of simple living makes room for rearticulation of desires and pleasures and how it restructures a hierarchy of what globalized consumer capitalism has positioned as objects of luxury. To illustrate the resistant potential inherent in the postulates of simple living, regardless of their frequent rhetorical excessiveness and pomposity, let us refer to one of the mini-chapters in *Inner Simplicity*, entitled "Learn to enjoy the silence." The revalorization of silence in a culture of polyphony of sounds and constant exposure to sounds as an immediate effect of social pressure to be active and efficient, is indicative of a desire to embrace these "other pleasures" which have been removed from the mainstream. Learning how to appreciate what Sara Maitland has called "the energy of silence"[58] may be interpreted as an attempt to "offer blueprints or projections of other possible futures, but in the sense of seeking to form desire, and to encourage a different structure of feeling and affective response to the world of material culture."[59] A retreat into silence "by creating as much quiet in your own space as possible" (*IS*, 21) suggests a need to disengage from the rhythms of everyday routines and a desire to pursue other pleasures.

Slow practices performed under the banner of slow living and the phenomenon of downshifting discussed above are representative of what might be described as holistic or systemic adaptations of the premises of alternative hedonism. The imperative of "doing things differently" reveals its versatility and permits one to create a system which challenges the dominant models of the good life by embracing various spheres of life, from work and leisure to material and spiritual aspects of the everyday life. When viewed from such a perspective, slowness emerges as a ground upon which alternative lifestyles may be erected and the "other" pleasures exposed. Interestingly, the recent popularity of slow narratives has created a variety of spaces where singular acts of celebration of the pleasures of slowness may be performed and represented. Not always or necessarily inscribed in wider systems of holistic lifestyle changes or motivated by ethical concerns, they illustrate the ways in which the cultural range of the present has been extended to articulate desires often suppressed by the logic of

---

58   Sara Maitland, *A Book of Silence* (London: Granta, 2009), p. 16.
59   Kate Soper, "The Other Pleasures of Post-Consumerism," *Soundings. A Journal of Politics and Culture*, Issue 35 (2007), p. 35, accessed 15 August 2013, http://www.lwbooks.co.uk/journals/soundings/issue/35.html

the culture of speed. Two different activities, cycling and the slow contemplation of works of art, may on one hand exemplify individual attempts to negotiate discontents of accelerated lifestyles, and on the other illustrate the expansion of slow philosophy and its celebration of alternatively defined pleasures.

While the car became a symbol of the new pleasures of speed in modernism, the bicycle has recently become a prop of the slow, sustainable and pleasurable life. To study cultural history and representations of the bicycle, "poised intriguingly between slowness and speed,"[60] is to study speed-oriented dreams and anxieties and to confront the tensions and contradictions of which the bicycle culture is not free. Interestingly, the bicycle in its early, pre-automobile days, was for a while a prop of the culture of speed and introduced man to a whole range of new experiences. In a manner that some time later would be repeated by the car, the bicycle freed cyclists from the dictates of train timetables and gave them access to "the 'pleasure in unfettered mobility.'"[61] Moreover, it was invested with the subversive potential to liberate men (and women) from the oppressive character of cultural norms, which were the guardians of morality and decency. A cyclist, immersed in the sensuous experience of the pleasures of speed, soon became a symbol of freedom of mobility and sexual expression, as illustrated in a novel by Maurice Leblanc, *Voici des ailes!* (1898). Stephen Kern's analysis of the novel's cultural representations of the bicycle calls up the image of a technological achievement which turned out to have a power to "bring the sexual, social and spatial liberation" to the protagonists, who discovered their real passions and desires while taking a bicycle tour,[62] and contradicts John Urry's reference to the bicycle as "humble."[63]

Today, the use of the bicycle may be invested with equally subversive implications and assume a less humble nature. In the discourse of slow life, which has revived the cycling culture and has built a chain of new cultural meanings around the bicycle, the pleasures of cycling are often represented as transgressive, since to use a bicycle when others drive cars means to challenge "dominant norms of convenience, of visible status and of speed."[64] Bicycle riding may also be seen as a manifestation of one's commitment to sustainable transportation and provide a cyclist, this model slow traveller, with satisfaction caused by a sense of agency.

---

60 Ryle, Soper, "Alternative Hedonism: The World by Bicycle," p. 99.
61 John Urry, *Mobilities* (Cambridge: Polity Press, 2007), pp. 112–113.
62 Stephen Kern, *The Culture of Time and Space 1880–1918* (Cambridge, MA.: Harvard University Press, 1983), p. 111.
63 Urry, *Mobilities*, p. 112.
64 Ryle, Soper, "Alternative Hedonism: The World by Bicycle," p. 95.

In this way a sense of pleasure is derived from the fact that "[…] the individual acts with an eye to the collective impact of aggregated private acts of affluent consumption for consumers themselves, and takes measures to avoid contributing to it."[65] A decision to ride a bicycle rather than drive a car may reflect a sense of responsibility for the ecological wellbeing of the environment, and manifest one's commitment to the slow cause by showing participation in a form of "resistance to one-dimensional narratives of progress."[66]

The pleasures of slowness may take various forms and may be experienced in various areas and contexts of life. The enthusiastic response with which Petrini's ideas have been met all over the world evidences the vitality with which they cut across various spheres of cultural practice and social demand for a proposal that would allow for a negotiation of existing models of the good life. Petrini says that "to live pleasurably, we need to broaden the range of things that give us pleasure, and that means learning to choose differently, even to live differently."[67] As we have seen, "choosing and living differently" in the context of the temporality of pleasure introduces one to the space of mindful, decelerated production and consumption and a slow celebration of the pleasures they bring. However, it does not have to be restricted to consumption (or reduction of consumption) of material goods only. What Petrini says in "Pleasure Denied, Pleasure Rediscovered" has been adopted as a starting point to reflect on the possibility to create alternative conditions of engagement with contemporary models of the experience of art. Slow Art Day, started by Phil Terry in 2009, was inspired by his own experience of pleasure derived from immersion in the slow study of one work of art in a manner that challenged popular modes of art experience. Jeffrey K. Smith and Lisa F. Smith's study of the amount of time spent viewing works of art by a group of 150 visitors to the Metropolitan Museum of Art showed that the median time spent before each work was 17 seconds.[68] Terry's idea was to create and publicize an alternative which would awake the desire to reduce the hasty consumption of works of art by rediscovering the pleasures of slow, mindful and pleasurable contemplation.

The idea of Slow Art Day has been popularized all over the world. Organized in April, Slow Art Day in 2014 was hosted in 225 venues in the United States,

---

65 Soper, "Re-thinking the 'Good Life'…," p. 211.
66 Ryle, Soper, "Alternative Hedonism: The World by Bicycle," p. 94.
67 Petrini, *Slow Food*, p. 21.
68 Trent Morse, "Slow Down, You Look Too Fast," *ARTnews*, accessed 26 April 2014, http://www.artnews.com/2011/04/01/slow-down-you-look-too-fast/. For the original study see: Jeffrey K. Smith, Lisa F. Smith, "Spending Time on Art," *Empirical Studies of the Arts*, Vol. 19, Nr 2 (2001).

Great Britain, the Netherlands, Italy, Greece, France and Australia, among other countries.[69] Those who are interested sign up at a local museum or gallery, attend on a day marked as Slow Art Day and look at pieces of art already selected by a host for 5 to 10 minutes, to discuss them later with other participants of the event. The "feats of prolonged attention,"[70] as Yve-Alain Bois describes slow contemplation of works of art, require a deceleration of gaze and mindfulness which accelerated lifestyles rarely allow for. According to Bois, some paintings, more than others, require suspended perception and "deflect the ever-growing demand for speedy consumption"[71] by embracing "the duration of perception into their aesthetic structure."[72] For that reason they are "among the strongest agents of resistance against the growing desensitisation of human subjectivity promoted by the so-called digital revolution."[73] Slow perception of art, performed individually or under the banner of Slow Art Day, is informed by a desire not to perceive the world as a "blur" and to restore the type of pleasure that contemporary culture tends to annul and that both slow and simple living defend.

Conclusions

The changing definitions of pleasure tell a story of moral, psychological, political and cultural shifts which have marked the character of everyday life. The way pleasure is conceptualized and represented reflects a system of values constructed and practised in particular cultural contexts. When seen from the velocentric perspective, pleasure reveals its embrace of a sensation of speed, the origins of which go back to modernism which created the car, a tool of individual control of speed, and made the experience of speed available to almost everyone and everywhere. Since that moment speed and pleasure have consolidated and started to reflect culture's turns and fluctuations.

A growing sense of dissatisfaction with the character of life in what seems to be a more and more dystopian society has brought about a return to pleasures and models of the good life which not so long ago were commonly associated either with the margins of cultural practices or with one's inability to meet the

---

69   In Poland in 2014 Slow Art Day took place in the Wrocław Contemporary Art Museum.
70   Yve-Alain Bois, "Slow (Fast) Modern," in *Speed Limits*, ed. Jeffrey T. Schnapp (Milan: Skira, 2009),
71   Bois, "Slow (Fast) Modern," p. 125.
72   Bois, "Slow (Fast) Modern," p. 125.
73   Bois, "Slow (Fast) Modern," p. 126.

requirements of high modernity and its rate of development. For Soper it is obvious that "[t]he long-standing cross-party consensus on what counts as 'high' living standards needs to be challenged through provision of a new 'political imaginary.'"[74] Slow living and practices inspired by its philosophy may be interpreted as attempts to create a space for the articulation and practice of alternative models of the good life and pleasure, while the rhetoric of slow living creates a reservoir of images and emblems of new definitions of wellbeing which, often invested with resistant undertones, indicate the possibility of living differently in an increasingly homogenized and globalized culture.

Paraphrasing Jeffrey T. Schnapp, who once observed that "slowness in and of itself is no guarantee of quality,"[75] it would be true to say that slowness itself is no guarantee of pleasure. But it may be a good start. The discourse of slow life rediscovers the pleasures of "gazing at God's windows" and by implicating these pleasures in the network of meanings and signifying practices of an accelerated culture reveals their potential to trigger social change and serve as models of alternative lifestyles, the demand for which has become a recurring theme in critical reflections about the conditions of the present, temporality, sustainability and the good life.

---

74 Soper, "Humanities...,".
75 Jeffrey T. Schnapp, "Fast (Slow) Modern," in *Speed Limits*, ed. Jeffrey T. Schnapp (Milan: Skira, 2009), p. 34.

Part Three

The Other Speed: Cultural Practices and Representations

Chapter Five: New Territoriality in the Age of Deterritorialization

> The terrain of a king's everyday life is not his country but his court.
> Agnes Heller, *Everyday Life*
>
> Space is the ongoing possibility of a different inhabitation.
> Elizabeth Grosz, *Architecture from the Outside:*
> *Essays on Virtual and Real Space*

In 1913 Gino Severini observed that "[s]peed has given us a new conception of space and time, and consequently of life itself."[1] The Italian Futurists sketched prophetically the contours of a world in which distance would be annulled by speed and where the solitude and nostalgia caused by spatial separation would be eliminated from the spectrum of experience. Almost a century later, in 2012, Olga Tokarczuk, a Polish novelist, published *Moment niedźwiedzia*, in part a travelogue containing notes and observations made in places she had visited. In a subchapter entitled "Maps of fear or *nil desperandum*" she reflects on her own spatial experience of Amsterdam, which she contrasts with "many other cities – sprawling, too big, with wide streets, built for the wind rather than for man," where walking is often perceived as a curiosity, where streets are sites of a permament struggle to survive and where people "just run their errands, rush, pass by."[2] While other cities may be shaped by "the new aesthetics of speed,"[3] enthusiastically praised by F. T. Marinetti, Amsterdam for Tokarczuk wears a size HS, a *homo sapiens* size,[4] which encourages encounters and generates a sense of safety, and where the city movement flows "graciously as a dance, unhurriedly and politely."[5] Putting aside what might strike one as a mythologizing of Amsterdam, which as a popular tourist destination often gives a radically different impression to those who only come to visit it for a day or two, Tokarczuk's urban sketch seems to be

---

1 Gino Severini, "Plastic Analogies of Dynamism: Futurist Manifesto," in *Futurism. An Anthology*, eds. Lawrence Rainey, Christine Poggi, Laura Wittman (New Haven: Yale University Press, 2009), p. 168.
2 Olga Tokarczuk, *Moment niedźwiedzia* (Warszawa: Wydawnictwo Krytyki Politycznej, 2012), p. 89. Translation mine.
3 F.T. Marinetti, "Le Futurism," in *Futurism. An Anthology*, eds. Lawrence Rainey, Christine Poggi, Laura Wittman (New Haven: Yale University Press, 2009), p. 99.
4 Tokarczuk, *Moment niedźwiedzia*, p. 87.
5 Tokarczuk, *Moment niedźwiedzia*, p. 86.

framed by the conviction, also informing the analytical angle in this chapter, that the space we live in and the way we use it are more than reflections of a worldview and the needs of those who inhabit the space. Spatial relations and space organization determine the character of social performance by creating conditions which promote desirable forms of social conduct and exclude those which might threaten the stability of the existing social, economic or cultural order.

The focus of this chapter is on changes occurring in the ways private and public spaces have been conceptualized, represented and reconstructed in the context of a growing awareness of the consequences of technological acceleration and the recent valorization of slowing down. I shall study two types of "slow" spaces. The first type includes spaces whose potential for slowness has been rediscovered, by both practitioners and theoreticians of slow living, under the influence of Carlo Petrini's call to oppose the fast life. These are private spaces (*Slow Home*) and public spaces (*Slow Cities*) which have become sites of negotiations between various temporalities. The choice of the second type of slow spaces has been inspired by a desire to examine the possibility of slowing down in an environment which by definition is hostile to slowness, i.e. the metropolis.[6] Supported by Nick Osbaldison's claim that to talk about cultural aspects of slowness (and, in the case of the present analysis, its spatial contexts) we need to "avoid discussing the culture of the slow through only explicitly named 'slow' movements,"[7] I will look at the ways in which a need to integrate spaces of different speeds shapes and reconstructs spatial practices and influences social life in large cities whose dynamics feed on rushing and fast movement. A focus on cultural practices which, although not performed under the banner of slow life, are nevertheless motivated by a desire to confront the consequences of acceleration in everyday life, allows me to expand what might otherwise be seen as a rather limited analytical scope. Viewed from such a broader perspective the quest for slowness may be represented not so much as part of the praxis promoted by social movements and popular fashion but rather as a cultural principle which is activated in response to acceleration when the latter reaches a critical degree. In the light of the analysis of the so-called "copenhagization"

---

6   While in this chapter I look at the possibility of slowing down in the city from the perspective of reorganization of urban spaces (Copenhagen), a part of Chapter Six explores slowness in the metropolis in the context of the slow tourist experience and its representations (London).

7   Nick Osbaldison, "Consuming Space Slowly: Reflections on Authenticity, Place and Self," in *Culture of the Slow. Social Deceleration in an Accelerated World*, ed. Nick Osbaldiston (Basingstoke: Palgrave Macmillan, 2013), p. 89.

and of the relationship between speed and community life in the city, a quest for spaces of slowness may be construed as a wish to reclaim pleasures long forgotten in the culture of speed.

All the phenomena discussed in this chapter are motivated by a concern for the character of the spaces we inhabit and are informed by a conviction that access to spaces of deceleration is fundamental for sustainable living and that it may lead to a revitalization of places, local cultures and communities. With regard to a sense of responsibility for the place one inhabits, this chapter concentrates on forms of spatial organization and performance as instances of what I propose to call *new territoriality*. The notion is used throughout the chapter to register the rise of a new spatial consciousness, which manifests itself in attempts to negotiate the cultures of the flow and their discontents and to challenge the displeasures of deterritorialization. While, as William Gaudelli and Timothy Patterson have noted, "discourses about human agency typically ignore spatial dimensions,"[8] the concept of new territoriality highlights the way in which both individual and communal experience of the relations of domination and resistance may trigger social changes through spatial reorganizations.

1. New Territoriality

Modern spatial relations illustrate Paul Virillo's famous claims that "speed is the cause and not the effect, since it dilates time just as it contracts space [...]."[9] The development of digital technologies of communication have speeded up the contraction of space and have produced the supremacy of simultaneity over linearity, which, as John Urry emphasized, "means that identities may well be less place-based and more engendered through relations made and sustained on the move, in liminal 'interspaces' [...]."[10] In her study of the nature of contemporary spatial experience and its representations in audio-visual culture Barbara Kita noted that being "here and now" is no longer a guarantee of success because of the pressures produced by an imperative of instantaneous temporality. The fixity of the phrase "here and now" has been broken: "now" has been liberated from what had been considered the unavoidable company of "here." The phrase

---

8   William Gaudelli, Timothy Patterson, "It's *Just* Geography: Critical Geography and a Critique of Advanced Placement Human Geography," in *Geography and Social Justice in the Classroom*, ed. Todd W. Kenreich (New York: Routledge, 2013), p. 116.
9   Paul Virillo, *Negative Horizon. An Essay in Dromoscopy* (London: Continuum, 2007), p. 129.
10   John Urry, *Mobilities* (Cambridge: Polity, 2007), pp. 176–177.

"here and now" has evolved into "here, now and somewhere else."[11] Dynamic conquests of public spaces by the cultures of the flow – which prioritize mobility, non-places, "an understanding of the planet as a 'single place,' a singular 'global' space,"[12] and a separation of space from place through the suspension of physical, localized presence – have resulted in the intensification of deterritorialization and a decrease of the singularity and uniqueness of place.

Wendy Parkins and Geoffrey Craig place deterritorialization in the context of globalization, and their approach is marked by a focus on the exchange between local and global forces. Although they do not question the fact that globalization is one of the major factors responsible for the transformation of locality and a consequent waning of the specificity of place, they do not support the view that in the age of global influences the local has to be annulled under the influence of globalizing forces. On the contrary, they observe that "the material realities of physical embodiment and the ongoing need for situated political, social and economic relations mean that the local persists as a significant site in global culture."[13] The rehabilitation of "here," which the discourse of slow living underscores, may thus be viewed as an answer to deterritorialization, which, as Parkins and Craig observe, "paradoxically can contribute to a revitalization of the site of 'the local.'"[14]

The exchange between the global and the local in the contexts of spatial dimensions of slow living is foregrounded by the concept of *terroir*. Used traditionally by food experts, it refers to "the specificity of the place that stems from its traditions, the uniqueness of local food culture and regional produce, that resides in the landscape, soil and climate, as well as the types of food grown, the farming techniques, and the cultural contexts that informs food preparation and consumption."[15] Important in the philosophy of the Slow Food movement, the concept of *terroir*, "this transmission of the taste of a parcel of land,"[16] exposes the way in which global forces, with their stress on homogeneity and erasure of difference, may activate place-specific performance and promote accentuation of a specificity of place. "At its simplest the glass of wine that we enjoy is a reflection

---

11   Barbara Kita, *Między przestrzeniami. O kulturze nowych mediów* (Kraków: Rabid, 2003), p. 9. Translation mine.
12   Wendy Parkins, Geoffrey Craig, *Slow Living* (Oxford: Berg, 2006), p. 70.
13   Parkins, Craig, *Slow Living*, p. 71.
14   Parkins, Craig, *Slow Living*, p. 69.
15   Parkins, Craig, *Slow Living*, p. 76.
16   Robert A. Davidson, "*Terroir* and Catalonia," *Journal of Catalan Studies* (2007), p. 41.

of the environment that it was grown in [...],"[17] says Janet Lymburn. The idea of *terroir* is an exemplification of the fact that the local may unite individual and global perspective by escaping the either/or logic of current models of representation and experience of locality: either "creeping global sameness" or "hyperindividualised difference."[18] The case of the Slow Food movement and its revalorization of locality, coupled with the movement's readiness to engage in a dialogic relation with global forces, is indicative of what Phillip Crang and Peter Jackson identify as "the interpenetration of the global and the local."[19] Their analysis of geographies of consumption corresponds with Parkins and Craig's reading of the meaning of the local in the context of the deterritorialization brought about by global forces. They claim that "[g]lobalization instigates a paradoxical process whereby a trend towards increasing global unicity in turn provides the means by which expressions of difference are more keenly identified."[20] *Terroir*, with its connotations of the uniqueness of a place and its history and agricultural and culinary traditions, prioritizes neither cultural fixity nor invariable attachment to the ideas of permanence. Rather, tradition is approached as a point of departure from which to (re)establish relations with the globalized present and modify them.

While the concept of *terroir* reveals the dynamics of interchanges both between the local and the global and between the past and the present, it seems viable to place it in a broader context which will allow us to read it as *one* of many manifestations of a recent return to a sense of place. Convinced that Parkins and Craig are right when they claim that "deterritorialization facilitates a cosmopolitan disposition that in turn offers the possibilities of 'reterritorialization,'"[21] throughout this chapter I will use the concept of new territoriality as an umbrella term to embrace a wide spectrum of types of spatial conduct which, although diversified, are still emblematic of a desire to negotiate actively the culture of the flow and its discontents. A study of new territoriality and its various forms from the velocentric perspective highlights the role which speed and slowness play in reconceptualization of the meanings of places and spaces and the ways in which revalorization of slowness helps invest them with new senses.

---

17 Janet Lymburn, quoted by Geoff Andrews in *The Slow Food Story. Politics and Pleasure* (Montreal: McGill-Queen's University Press, 2008), p. 131.
18 Philip Crang, Peter Jackson, "Geographies of Consumption," in *British Cultural Studies*, eds. David Morely, Kevin Robins (Oxford: Oxford University Press, 2001), p. 330.
19 Crang, Jackson, "Geographies of Consumption," p. 330.
20 Parkins, Craig, *Slow Living*, p. 71.
21 Parkins, Craig, *Slow Living*, p. 71.

Conceptually, new territoriality may be seen as situated at the intersection of two types of "metaphysical ways of viewing the world,"[22] which Tim Creswell identified in his study of Western mobility. Inspired by Liisa Malkki's anthropological analysis of mobility of refugees, in which she introduced the concept of "sedentarist metaphysics"[23] to refer to a link between a dominant worldview and a conceptualization of cultures as rooted and fixed, Creswell described two major models of an "understandings of mobility, spatial order, and place"[24]: nomadic and sedentarist metaphysics. While sedentarist metaphysics is defined by its concern with place, a sense of belonging and rootedness, nomadic metaphysics "puts mobility first, has little time for notions of attachment to place and revels in notions of flow, flux, and dynamism."[25] Creswell's analysis, carried out in areas as diverse as photography, airport culture, law and immigration, reveals the ways in which the signifying processes operating within these areas come to be informed by a sedentarist and/or nomadic metaphysics and helps demonstrate that, as Creswell is careful to note, "mobility, and the meanings given to it, permeates [sic] modern culture and society in the Western world."[26] Both types of metaphysics entail various representations and understandings of space and place and initiate different social and linguistic practices. The organization of space and spatial relations are subject to conceptualizations and (re)arrangements informed by modes of thinking about space which are suspended between valorization of place and a sense of limitations it may create.

Placing new territoriality within contexts of the metaphysics of mobility highlights the ways in which culture has recently modified such a bipolar understanding of mobility and the meanings of place by making the exchanges between the local and the global a part of everyday experience. While it is true to say that globalization and unprecedented mobility are responsible for the degradation of the meaning of place, they may also set in motion a revitalization of locality, a process which has recently gained more and more impetus. Taking place under the banner of resistance to acceleration and globalization, revitalization of

---

22   Tim Creswell, *On the Move. Mobility in the Modern Western World* (New York: Routledge, 2006), p. 26.
23   Liisa Malkki, quoted by Creswell in *On the Move*, p. 26. See also: Liisa Malkki, "National Geographic: The Rooting of Peoples and the Territorialization of National Identity among Scholars and Refugees," *Cultural Anthropology*, Vol. 7, No. 1 (1992).
24   Creswell, *On the Move*, p. 26
25   Creswell, *On the Move*, p. 26.
26   Creswell, *On the Move*, p. 25.

locality can enforce a shift towards the notion of place as a site of confrontation with those practices that dissolve the sense of territorial identification.

New territories, these reclaimed places, may be understood as a form of human organization of space informed on one hand by experience of an attachment rooted in a sense of co-responsibility for the character of place, which is seen as a "field of care,"[27] and on the other, by a desire to counteract the displeasures of the authoritarian rule of the culture of non-places. Seen in this light, new territories, which mark the rise of a new, post-deterritorialization dynamics, are not about power but about agency. New territoriality, experienced individually as an intimate relation with public and private spaces and collectively as a social situation, rests on two premises. Firstly, it is an effect of contextualization of one's spatial relation with regard to dynamic exchanges between the private and the public, highlighted by many contemporary discourses (the discourse of sustainability being a case in point). Secondly, what activates practices involved in the articulation of new territoriality is a desire to return to a sense of place, "to a heightened appreciation of regard for a specificity of experience."[28] New territoriality embraces a wide range of spatial practices exemplifying a compelling tension between various temporalities which reveals a growing desire to challenge the assumptions of the culture of the flow and negotiate its consequences.

2. Territories of the Slow Self: Slow Home and Slow City

Popularization of the idea of "slow spaces" has never been motivated by a drive towards "the establishment of alternative spaces"[29] which would function in a secure isolation from spaces the character of which has been dominated by speed. Rather, this idea has always been focused on the possibility of a "transformation of the spaces we currently inhabit"[30] in a way which would facilitate an integration of spaces of high and low speeds. That said, it is clear that this dimension of deceleration cannot be realized without the rise of a new spatial sensitivity, the presence of which is becoming more and more discernible in both private and public spheres. Introduction of the concept of deceleration into present cultural debates has increased awareness of a need to create, mark and protect those spaces which give man a chance to liberate himself, whether only for a moment or for longer, from the dictatorship of the nervous rhythms of the everyday.

---

27 Creswell, *On the Move*, p. 31.
28 Davidson, "*Terroir* and Catalonia," pp. 41–42.
29 Parkins, Craig, *Slow Living*, p. 62.
30 Parkins, Craig, *Slow Living*, p. 62.

This new spatial awareness activates a mode of thinking about space which conceptualizes deceleration as a value in public spaces and which is oriented towards reclamation of spaces appropriated by the culture of speed.

Recently, the spatial dimensions of slowness have been articulated in both private and public contexts. The private character of spaces of deceleration is most often associated with the space of home, which is traditionally represented as an oasis of intimacy.[31] In the conditions of modern mobility, flow and rush, the space of home is influenced by reflections on relations between slowing down and domesticity. The most common representation of home is that of a shelter which protects its inhabitants from the oppressive outside world and which is a bastion of slow time, relaxation and calm, a space "designated to insulate people from the perceived ravages of public life."[32] Interestingly, this somehow simplifying and reductionist bias of the bipolar logic of representation of domestic spaces has been recently placed in the spotlight of critical reflections under the influence of the philosophy of slow living. The oppositional character of a rush-vs.-slowness juxtaposition, which reflects a dichotomic pattern of the inside/outside, has been subjected to redefinition. Seen from the perspective of slow living the space of home is no longer approached as a space of withdrawal, which may facilitate social isolation, but on the contrary, it is represented as a space of vital interaction with the outside world. In slow living, Parkins and Craig emphasize, the space of home is the site of practices which are then transferred to the world outside; the character of domestic space is no longer shaped by its opposition to the outside world but rather through interaction with it. They claim that such a representation of home is "defined fundamentally through its articulation with broader social contexts and through the desire for it to impact upon society [...]. Practices grounded in the home space, that is, can be means of engagement with the world rather than a retreat from it."[33]

Unsurprisingly, the concept of slow home has been adopted and popularized by a commercial company which has used it to propose an architectural response to "the poor design practices that pervade the mass housing industry."[34] The Slow Home movement, created in the wake of a recent explosion of interest in slowness, was started in 2006 by John Brown, Matthew North and Carina van

---

31 Parkins, Craig, *Slow Living*, p. 66.
32 Parkins, Craig, *Slow Living*, p. 66.
33 Parkins, Craig, *Slow Living*, p. 66.
34 "Our Philosophy," Slow Home Studio, accessed 19 November 2013, http://slowhomestudio.com/our-philosophy/. Throughout this chapter the website will be reffered to as *SHS*.

Olm with the purpose of promoting a "more thoughtful approach to residential design that improves the quality of our daily lives and reduces our impact on the environment" (*SHS*). It is typical of the present-day reversal of traditional connotations that the fast/slow binary generated over the last two hundred years, that the founders of the Slow Home movement use the term "slow" to create a division of houses we inhabit into "fast" and "slow" ones. According to the philosophy of the Slow Home

> a fast house is a standardised, mass-produced commodity that has been designed to attract our attention, ignite our desire, and give the illusion of value as much if not more than it's been designed as a place to live. This lack of attention to the fundamentals of good design makes a fast house difficult to live in and hard on the environment (*SHS*).

The paradox behind this representation is that while fast houses are represented as commodities and mass products, slow houses are also commodities in the sense that what is purchased is not only a sustainable space to live but also, if we follow Pierre Bourdieu in this respect, a certain form of social capital. Slow Home is a "materialisation of class taste,"[35] since the issue of its financial cost is significant. This brings us back to the question of the pleasures of slowness, which are represented by slowness activists as being available to everyone. While cycling or travelling by slow train may be examples of such "democratic" pleasures, the pleasures of living is a "slow," sustainable home are available only to the affluent classes of Western society.

While the Slow Home Studio's rhetoric may be approached as an example of commercialization of the slow logo and the middle class appetites for the latest trends, it may also be read as a manifestation of a growing spatial sensibility emblematic of the recent attempts to conflate good, sustainable life with slow life. The owners of the company explain: "[w]e believe that homes are too emotionally significant, have too large an environmental footprint, and represent too significant a financial investment for this kind of institutionalized bad design to continue unchecked" (*SHS*). In the light of Parkins and Craig's analysis of slow temporality in private space, this mode of thinking about home as a site of sustainable living reflects a turn towards domestic practices as a means of interacting with the world and negotiating various social issues and concerns. The "Slow Home Test" prepared by the Slow Home Studio to facilitate an evaluation of the quality of design includes three sections: "The House in the World," "The

---

35 Pierre Bourdieu, *Distinction: A Social Critique of the Judgement of Taste*, trans. Richard Nice (London: Routledge, 1984), p. 23.

House as a Whole" and "Room by Room."[36] While the last two sections focus on the design of the interior of the house, with special emphasis paid to the issues of light, circulation, size and storage, the first section expresses awareness of the significance of the exchange between the space of the house and the world outside the house. The "slowness" of the home and its sustainability are evaluated with reference to its location and orientation, and the evaluation stresses, respectively, the walkability of the area and the building's potential to make use of the sun as the source of energy. The space of the house is no longer a private space designed to satisfy the needs of its inhabitants only, but becomes a territory that allows to "retain an ethical and political disposition that is grounded in an awareness of our fundamental relationships to the specificity of place, the land, its produce and each other."[37] The inhabiting of one's private space and the world have common denominators: a sense of co-responsibility for the natural environment and a sense of agency, which gives one a feeling of participation in practices that shape the character of everyday life. A sense of "ownership" does not stop at the edge of domestic space but is extended to embrace local and global environments. Although on a different scale and to a different extent, sustainable living at home and in the world demand similar attentiveness and mindfulness.

The mode of thinking about space organization and co-responsibility for the ecological wellbeing of the environment that permeates the ideological frameworks of the Slow Home movement, may be examined as a new spatial awareness and a manifestation of new territoriality. For Parkins and Craig this new type of reconceptualization of spatial relations "requires particular spatial contexts as much as it requires a slow temporality,"[38] and it finds its most organized and institutionally supported form in the Slow Cities movement.

The Slow Cities network, the flagship product of the Slow Food movement and an international phenomenon, was started in 1999 in Italy by the mayors of four Italian towns: Bra, Orvieto, Positano and Greve-in-Chianti. By now the Cittaslow International Network consists of 187 towns (only those with less than 50,000 residents may join) in 28 countries all over the world, according to the list updated in April 2014.[39] The official website of the movement says

---

36 "Slow Home Studio Test," accessed 19 November 2013, http://slowhomestudio.com/wp-content/uploads/SlowHomeTest51.pdf
37 Parkins, Craig, *Slow Living*, p. 85.
38 Parkins, Craig, *Slow Living*, p. 63.
39 "Cittaslow List," accessed 26 April 2014, http://www.cittaslow.org/download/DocumentiUfficiali/CITTASLOW_LIST_april_2014_PDF.pdf. Slow cities may be

that "[…] living and managing a Slowcity is just a particular way of carrying on an ordinary life-style rather than today's trends."[40] The principles of the movement are based on the belief that cultural deceleration may trigger needed spatial reorganization and facilitate the implementation of changes which would influence such areas as communication, production, trade and transportation. The underlying conviction is that "an ordinary lifestyle" which slow cities make possible "will be more human, environmentally correct and sensible for the present and future generations."[41]

In a brief history of the Cittaslow movement we read:

> Municipalities which join the association are motivated by curious people of a recovered time, where man is still protagonist of the slow and healthy succession of seasons, respectful of citizens' health, the authenticity of products and good food, rich of fascinating craft traditions of valuable works of art, squares, theaters, shops, cafés, restaurants, places of the spirit and unspoiled landscapes, characterized by spontaneity of religious rites, respect of traditions through the joy of a slow and quiet living.[42]

The ideology of the movement is permeated by recognition of the role one's relation with space, especially public space, plays in the construction of the good life and by the conviction that the broadly understood protection of the local character of a place constitutes a prime value. The recognition of the significance of locality inspires resistance against globalization which "although representing an opportunity of exchange and circulation, has a tendency though, to flatten-out the differences and hide the features typical of the individual communities […]."[43] Today spatial practices favour "average models"[44] that belong to no one and thus create spaces that lack capacity to develop and maintain their own specific and unique identity. For the authors of the "Cittaslow International Charter" what is produced in "no man's lands" is mediocrity which, in a rhetorical manner

---

    found not only in European countries such as Austria, Belgium, Poland, Hungary, Sweden and Finland, but also in Australia, Canada, China, Japan, South Korea and the USA.

40   Cittaslow International, "Philosophy," accessed 12 April 2014, http://www.cittaslow.org/section/association/philosophy

41   Cittaslow International, "Philosophy."

42   "About Cittaslow Organization," accessed 1 May 2014, http://www.cittaslow.org/section/association

43   "Cittaslow International Charter," Attachement "A" to the Charter, p. 20, accessed 25 April 2014, http://www.cittaslow.net/download/DocumentiUfficiali/2009/newcharter%5B1%5D.pdf

44   "Cittaslow International Charter," p. 20.

typical of most "slow manifestoes," they contrast with an excellence born out of the unique character of locality and its cultural and historical heritage. The image of towns brought to life by mindful appropriation of reorganized spaces is constructed in opposition to that of homogenized towns whose identities are dissipated under the influence of globalized forces. The slow cities movement declares a desire to oppose the homogenizing practices of an age of acceleration by defending the uniqueness of particular places and by appealing to a possibility of a return to pleasures currently made unavailable by the logic of what Jeremy Rifkin called "the economy of speed,"[45] the central aspect of which has been quickness of production, consumption and distribution of goods and services.

The rhetoric of the Slow Cities movement is seductive since it appeals to the attractions of an alternative relation with space and accentuates not only "altruistic compassion and environmental concern, but also [...] the more self-regarding gratifications of consuming differently."[46] The images of slow living in slow cities revolve around the values of "places of the spirit and unspoilt landscapes" and thus suggest an alternative model of sustainable good life. Seen in the light of Kate Soper's alternative hedonism, the rise of the slow cities movement illustrates the energy and vitality of people's desire to live differently, "partly because of their concern about the long-term global consequences of current modes of consumption and partly because the existing ways are increasingly at odds with their own pleasures."[47] The pleasures of living in a slow city derive from, among other things, a sense of co-responsibility for the immediate environment, the promotion of a type of urban planning which encourages people to spend time in public spaces, a focus on natural and traditional food production, popularization of the ideas of sustainable living through the system of education, and the promotion of hospitality understood as "removing physical and cultural obstacles that may prejudice the full and widespread use of the city resources."[48]

An integral element of the philosophy behind the phenomenon consists in "foregrounding of the value of traditional ways of life and the historicity of the place."[49] A valorization of the historical dimension of place does not mean

---

45   Jeremy Rifkin, quoted by Robert Hassan, *Empires of Speed* (Leiden: Brill, 2009), p. 21.
46   Kate Soper, "Introduction: The Mainstreaming of Counter-Consumerist Concern," in *The Politics and Pleasures of Consuming Differently*, eds. Kate Soper, Martin Ryle, Lyn Thomas (Basingstoke: Palgrave Macmillan, 2009), p. 4.
47   Kate Soper, "Interview. An Alternative Hedonism," interviewed by Ted Benton, *Radical Philosophy* 92 (1998), p. 33.
48   "Cittaslow International Charter," p. 21.
49   Parkins, Craig, *Slow Living*, p. 77.

cutting it off from the present or from the past. On the contrary, the founding fathers of the Slow Cities movement are motivated by an urge to transform cities into places the shape of which is a function of tradition and modernity, locality and globality. Slow cities are a testimony to the fact that in an age of homogeneity and mass production the creation of spaces which promote sustainable growth may actively and creatively appropriate historic specificity of place, discover its unique character and reinforce a sense of individual and communal identity. The local is no longer "an appendage to 'the global,'"[50] but becomes a source of identification through the promotion of particularities which prevent the forming of the plaque of sameness responsible for homogenization of many towns around the world at the cost of the loss of their historical and cultural exceptionality.

The rise of the territories of the slow self, be they private or public, is an illustration of modern consumers' active search for a new conception of the good life. "Spaces of slowness" stress a new form of the relation between individuals and places, the popularization of which owes much to the ideas put forward by Carlo Petrini. This relation is represented as invested with a sense of co-responsibility for the character of place, a desire for sustainable growth, and a belief that reorganization of spatial design, both at home and in public space, may result in a quality life and help one confront the discontents caused by accelerated technological development.

3. "The Copenhagen Treatment": From Vehicular to Pedestrian Culture

The concepts of an alternative good life inspired by a revalorization of slowness assume neither a boycott of the spaces of fast life nor their eradication in the name of praise of slowness. This is so not only because such postulates would sound utopian, but also because slow spaces result from a desire to create a *balanced* sense of spatial experience which can only be possible when spaces of slowness function *next to* spaces of flow. The concept of good living in urban space is being redefined to meet the requirement of accessability to spaces marked by differentiated time regimes and to "multiplicities of 'slowness' and 'speed.'"[51] A demand for coexistence of areas of different temporalities is emblematic of what Parkins and Craig see as "fundamental public spatial requirements that are at least undermined by the contexts of 'fast life': the need

---

50 Geoff Andrews, *The Slow Food Story: Politics and Pleasure* (Montreal: McGill Queens University Press, 2008), p. 129.
51 Parkins, Craig, *Slow Living*, p. 73.

to disengage – however momentarily – from the flow of urban life for individual orientation, attention and contemplation [...]."[52] This brings to mind the quest for pleasures which have been increasingly repressed by the pace of life and by a character of spaces which this culture promotes: spaces of flow and mobility in which slowing down is conceptualized and represented as an obstacle to efficiency and time-saving

While Slow Home and the Slow Cities movement are extensions of the Slow Food movement, an examination of the spatial dimension of deceleration is by no means an exclusive product of the "slow revolution" initiated in Italy in the last decades of the 20th century. Though it is true to say that the Slow Food movement has inspired articulation of anxieties and concerns brought about by the modern culture of speed, it is also true to say that various models of spatial resistance against acceleration and its consequences have been constructed independently from the postulates of Slow Cities. When Nick Osbaldiston says that "the negotiation of space within the everyday has been limited in relation to slowness"[53] he is right in the sense that it has been only the recent popularization of slow ideas that has opened the doors for slow-oriented spatial and architectural practices on such a scale and with such publicity. However, right as he is, Osbaldiston should not make us ignore attempts to reclaim spaces of the flow which were successfully introduced long before the rise of the Slow Movement and which are illustrative of a potential to engage with urban spaces on different terms. These attempts illustrate acts of negotiation of various temporalities outside of social engagements associated with the Slow Food movement and may be placed in the broader context of what may be interpreted as culture's strategic resistance to acceleration.

The conceptual similarity between the postulates of the Slow Cities movement and the ideas popularized by Dutch architect Jan Gehl demonstrates the increasing vitality of "slow postulates" that gradually break into the consciousness of modern city dwellers. Motivated by a desire to protect spaces which allow one to slow down, the rehabilitation of slowness in urban spaces, later accentuated and promoted by the Slow Cities movement, started long before the rise of the movement itself. An example which shows the dynamics and logic of this process is the series of changes which have been introduced in Copenhagen since the 1960s, now commonly referred to as "copenhagization." As a result of changes made in Copenhagen over the course of almost five decades, the city earned the

---

52   Parkins, Craig, *Slow Living*, p. 74.
53   Osbaldiston, "Consuming Space Slowly ...," p. 72.

reputation of one of the world's most sustainable metropolises. When in 2006 Melbourne set out to create a network of European-style cycling lanes, the project was called "The Copenhagen Treatment"[54] to emphasize the role Copenhagen plays in a popularization of the idea of good urban spaces in cities all over the world. Copenhagen has established a model for the reconstruction of urban spaces, revitalization of the city centre and re-conquest of the space of the city from the oppressive grasp of the car culture.

The process of transformation of public spaces in Copenhagen started in 1962 when the old main street, Strøget, was closed to cars. The elimination of car traffic from the city centre was the aim of urbanists and city planners who meant to resist the tyranny of the car culture and reclaim the city for its residents. What initiated the spatial revolution in Copenhagen was the growing realization that a large portion of public space had been appropriated by cars. This, in turn, triggered the rise of a model of urban life dominated by automobility: "[u]p until 1962, all the streets in the city center were filled with car traffic, and all the squares were used as car parks. The post-war increase in car traffic meant rapidly deteriorating conditions for pedestrians in the city center."[55] The car-oriented urban planning created a vast range of "dead public spaces"[56] which encouraged fast mobility and discouraged slowing down.

Rebecca Solnit in her study of the cultural history of walking noted that "[i]n great cities, spaces as well as places are designed and built: walking, witnessing, being in public, are as much part of the design and purpose as is being inside to eat, sleep, make shoes or love or music."[57] However, she is also aware that the growth of modern cities redefined the character of public spaces, turning them into "merely the void between workplaces, shops, and dwellings."[58] The domination of car culture tyrannized urban space, relegating the pedestrian and the walker to the margins. The logic behind this was grasped by Richard Sennett, who wrote:

> Speed has thus become a means to the end of pure motion – we now measure urban spaces in terms of how easy it is to drive through them, to get out of them. The look of urban space enslaved to these powers of motion is necessarily neutral: the driver can

---

54 Clay Lucas, "Euro-style Bike Lanes Plan for City," *The Age*, accessed 23 April 2014, http://www.theage.com.au/news/national/eurostyle-bike-lanes-plan-for-city/2006/09/02/1156817151269.html
55 Jan Gehl, Lars Gemzøe, *Public Spaces, Public Life. Copenhagen*, trans. Karen Steenhard (Copenhagen: Narayana Press, 2004), p. 11.
56 Peter Freund, *The Ecology of the Automobile* (Montreal: Black Rose Books, 1993), p. 119.
57 Rebecca Solnit, *Wanderlust. A History of Walking* (London: Verso, 2001), p. 176.
58 Solnit, *Wanderlust*, p. 176.

drive safely only with the minimum of idiosyncratic distractions; to drive well requires standard signs, dividers, and drain sewers, and also streets emptied of street life apart from other drivers. As urban space becomes a mere function of motion, it thus becomes less stimulating in itself; the driver wants to go through the space, not to be aroused by it.[59]

The case of Copenhagen illustrates the change of a city centre from "a car-oriented to a people oriented city center."[60] The model of urban space proposed by Gehl has been marked by a departure from what Sennett recognized as a defining feature of modern urban space: that it became a means of motion. Sennett's representation of urban space as reduced to the function of movement and thus deprived of a stimulating value corresponds with Marc Auge's claim that today "the traveller is absolved of the need to stop or even look."[61] Informed by a growing displeasure brought on by a sense of disconnection from space and dissatisfaction with enforced modes of experiencing the world "in narcotic terms,"[62] a series of radical steps, designed to be implemented over a long period of time, was prepared for Copenhagen.

The starting point was to change the evaluation criteria of high-quality urban life. Whether the city was capable of creating conditions for good living ceased to be judged in terms of the ease and speed with which urban space could be crossed; what was taken into consideration was the quality and quantity of places in which to slow down and stop that the city could offer. The reclaiming of space conquered by the car turned out to be an indispensable condition for the creation of new, human-oriented space and marked the first stage of making the city wear a *"homo sapiens* size." The second stage of the reconstruction of Copenhagen's spatial character was informed by a conviction that quality of urban life depends not only on the elimination of cars from the centre but also on the city's capacity to initiate interactions between places and people. Jan Gehl and Lars Gemzøe claimed that "[o]ne way to judge quality in a city is not to look at how many people are walking, but to observe whether they are spending time in the city, standing about, looking at something, or sitting just enjoying the city, the scenery and the other people."[63] The return of the city's squares to its inhabitants

---

59 Richard Sennett, *Flesh and Stone. The Body and the City in Western Civilization* (New York: W.W. Norton & Company, 1994), pp. 17–18.
60 Gehl, Gemzøe, *Public Spaces…*, p. 6.
61 Marc Augé, *Non-Places. Introduction to an Anthropology of Supermodernity*, trans. John Howe (London: Verso, 1995), p. 97.
62 Sennett, *Flesh and Stone*, p. 18.
63 Gehl, Gemzøe, *Public Spaces…*, p. 59.

by banning car traffic and the transformation of public spaces to encourage stopping (benches, outdoor restaurants) may be read as a manifestation of what Andrzej Basista described as "nostalgia for streets and squares."[64] Basista posits a connection between the gradual disappearance of streets and squares from cities and modern changes of urban spatial organization which bring a gradual elimination of those spaces and places which encourage stopping, loitering, lingering or slowing down. Reclaiming spaces which had been taken over by the culture of speed and the subsequent restoration of squares in Copenhagen which were, not so long ago, occupied by parking lots can be read as "a critical response that is troubled by an intuition of the pleasures that are being directly occluded or denied."[65] The city's turn towards the "spaces of slowness" seems to be motivated by a conviction that a city dweller needs both to be able to move quickly and to slow down or stop. As we read in "The Living Streets Manifesto," "Living Streets need nooks and corners, benches and walls where people can pause and pass the time."[66] The city that opens a protective umbrella over its "slow spaces" diversifies its character and invites its inhabitants to co-create its character.

The changes in Copenhagen brought people to the streets. In spite of initial scepticism, fuelled by fears that the climate would not allow the popularization of a way of life typical of southern countries, the "Copenhagen Treatment" managed to instigate a mental change while bringing about a change of spatial organization: "[i]n 27 years in which this development has been followed, the number of people who spend time in streets and squares of the city center has increased 3,5 times. [...]. An interesting note is that over the same span of time, the total area of car-free streets and squares has increased 3,5 times."[67] The spatial changes in Copenhagen have helped revitalize the idea of public space as a cultural forum and the restoration of pedestrian areas has triggered the development of interactions which include informal activities such as the exchange of goods and viewpoints, and street entertainment and organized activities such

---

64 Andrzej Basista, "Spojrzenie z ukosa. Tęsknota za ulicami i placami," *Autoportret. Pismo o dobrej przestrzeni*, 1 (2004), pp. 28–29. Translation mine.
65 Soper, "Interview," p. 4.
66 "The Living Streets Manifesto," quoted by Martin Ryle and Kate Soper in "Alternative Hedonism: The World by Bicycle," in *Culture of the Slow. Social Deceleration in an Accelerated World*, ed. Nick Osbaldiston (London: Palgrave Macmillan, 2013), p. 100.
67 Gehl, Gemzøe, *Public Spaces* ..., p. 59.

as festivals.[68] The living streets and squares are a manifestation of "spatial consciousness"[69] which recently has been responsible for many changes in the organization of urban spaces aimed at creating more balanced and sustainable modes of living in towns and cities.

4. "Between the Buildings": Renewing Community

In *Ecological Literacy* David W. Orr referred to Henry David Thoreau's famous retreat to a wooden house by Walden Pond, saying that while living there Thoreau "revealed something of the potential lying untapped in the commonplace, in our own places, in ourselves, and the relation between all three."[70] Encouraged by Thoreau's readiness to acknowledge the role of one's (common) place in the process of identity formation, Orr examined various models of human inhabitation to come to the conclusion that "[i]t [the planet] needs people who live well in their places."[71] The idea of "living well" in one's place points towards a relation of a person to a place that consists not so much in possession – this place is mine and nobody else's – but in the formation of an emotional connection with it, which suggests that a place is a site of personal meanings. The (re)construction of the relation between the commonplace, one's own territory, and oneself, informed by a sense of co-responsibility for the character of place, turns out to be the foundation of the idea of living well in one's place and may be interpreted as the condition *sine qua non* of new territoriality.

The turn towards a new definition of "good places" and the concept of "living well" in one's place seems to be embedded in the conviction, recently more often and more clearly articulated in both popular and professional debates about space, that a responsibility for a place we inhabit is a form of self-responsibility and that a place and its inhabitant are engaged in a mutual process of constitution.[72] Self-responsibility as it is manifested in one's relation with place displays a certain conceptual similarity to the idea of self-interest highlighted by alternative hedonism and its orientation towards social change triggered by a desire

---

68   Gehl, Gemzøe, *Public Spaces* ..., p. 68.
69   Parkins, Craig, *Slow Living*, p. 63.
70   David W. Orr, *Ecological Literacy: Education and the Transition to a Postmodern World* (Albany: State University of New York Press, 1992), p. 125.
71   David W. Orr, *Earth in Mind: On Education, Environment, and the Human Prospect* (Washington: Island Press, 2004), p. 12.
72   Hanna Buczyńska-Garewicz, *Miejsca, strony, okolice. Przyczynek do fenomenologii przestrzeni* (Kraków: Universitas, 2006), p. 39.

to restore pleasures which have been marginalized by the culture of speed. Depersonalization of place, which often brings about spiritual homelessness,[73] can be reversed by developing a sense of responsibility for one's territory, which may become a point of departure to renew community and create space for a shared communal experience.

The disintegration of communities is a complex process encouraged by urbanization, a culture of speed and the development of modern communication technologies. The rise of the virtual world has further redefined community by freeing it from the spatial dimension, which was traditionally perceived as essential for a community to exist. In the "real" world (we use the distinction between the virtual and the real world only for the sake of clarity since it seems no longer reasonable to separate two spheres of life that have become so closely intertwined) the most available form of the experience of communality is provided through a "togetherness of consumption," which for Tony Judt is an "impoverished view of community,"[74] or by gated communities, the ghettoization of which constitutes the most radical form of spatial separation.

A recent turn towards deceleration and its powers to reconceptualize the meanings and values of cultural practices has played a crucial role in reviving interests in communities and in the ways in which they form and influence the quality of everyday life as sites of social, political and cultural exchanges between their members. Cecile Andrews observed:

> Community is so vital to a fulfilled life, yet building community is at the bottom of consumer society's list. As we run short of time, community is where we cut back. Why? First, we don't realize its importance. We don't realize what an incredible impact it can have on our lives. Further, building community is not something that yields instant rewards. The rewards come from staying in for the long run, not a value our society cherishes. Next, I think that because we have experienced community so rarely, we don't really even know what it is. How can we value it if we've never experienced it? Unless we develop a vision of what community can be we cannot move to create it. Finally, our community-building skills, in fact so many of our people skills, are stunted due to neglect.[75]

Andrew's engagement in the reconstruction of local communities has its source in the belief, often emphasized in public debates, that community-oriented living is a value which has been pushed to the margins of social life by a culture that, in

---

73   Buczyńska-Garewicz, *Miejsca, strony, okolice*, p. 39.
74   Tony Judt, *The Memory Chalet* (London: Vintage Books, 2011), p. 32.
75   Cecile Andrews, *The Slow is Beautiful. New Visions of Community, Leisure and Joie de Vivre* (Gabriola Island: New Society Publishers, 2006), pp. 174–175.

pursuit of maximization of profits and instantaneous satisfaction of needs, conceptualizes it either as a relict of the past or as a luxury only few can afford. The issue of community and its influence on the ideas of "living well" has emerged as one of the fundamental aspects of recent spatial practices performed under the slogan of slow living. Today a growing number of cities are ready to acknowledge a need to recreate those elements of social and individual life which have been wiped out or annulled, a sense of communal experience being one of them.

Ray Oldenburg observed that "[l]ife without community has produced, for many, a life style consisting mainly of a home-to-work-and-back-again shuttle. Social well-being and psychological health depend upon community."[76] His analysis of the erosion of communal life in modern cities led him to propose that what is indispensable for the recreation of community, in conditions whose character is marked by urban anonymity, is "the third place":

> Most needed are those 'third places' which lend a public balance to the increased privatization of home life. Third places are nothing more than informal public gathering places. The phrase 'third places' derives from considering our homes to be the 'first' places in our lives, and our work places the 'second.'[77]

Oldenburg's idea of the third place corresponds with Gehl's analysis of "life between buildings," to cite the title of one of his works. For Gehl, the city dwellers' activity in public places illustrates the ways in which physical environment shapes their relation with space they inhabit:

> Although the physical framework does not have a direct influence on the quality, content, and intensity of social contacts, architects and planners can affect the possibilities for meeting, seeing, and hearing people – possibilities that both take on a quality of their own and become important as background and starting point for other forms of contact.[78]

Highly critical of the faults of modernist urban planning, Gehl describes the relation between man and the city from the perspective that public space is a chance for the real existence of the city and a site of urban interactions. His analysis forms an important point of departure for a better understanding of the principles of the use of public spaces and consequently may lead to the improvement of the quality of life "between buildings."

---

76 Ray Oldenburg, "Our Vanishing 'Third Places,'" Planners Web, accessed 23 April 2014, http://plannersweb.com/1997/01/our-vanishing-third-places/
77 Oldenburg, "Our Vanishing 'Third Places.'"
78 Jan Gehl, *Life Between Buildings: Using Public Space*, trans. Jo Koch (Copenhagen: Arkitektens Forlag), p. 13.

Gehl's analysis distinguishes three major types of behaviour in public space. These are: necessary, optional and social activities. Necessary activities are carried out in order to realize obligatory aims – they include walking to work or school, shopping, performing one's professional obligations, such as delivering the post or distributing leaflets, street cleaning, etc. Optional activities take place when a man wishes to do them, when he can afford time needed to do them and when there are favourable conditions for their realization (e.g. weather). They do not result from a sense of obligation since they are motivated by one's own needs and desires. The third type of activities are social ones which depend upon the presence of other people in public spaces and which include both active and passive contacts. Active contacts consist in greetings, conversations and children's games, while passive contacts include watching and listening. The relation between the three types consists in the fact that the better the conditions for necessary and optional activities, the better are the social activities that are constructed and maintained.[79]

Such an approach to the life between buildings has its roots in a concern with the quality of life in cities, which has been subjected to deterioration caused by the territorial and demographic expansion of cities, and the acceleration of life, to such an extent that meaningful human encounters become impossible. The size of the cities and the speed of urban life generate and reinforce the dominance of passive social activities, while in slow spaces, such as the ones which slow cities recreate, active social contacts tend to dominate in public spaces, which there are oriented not towards facilitation of flow but rather of encounters. Gehl's point is that planning decisions can influence the forms of human behaviour and social interactions among the public. What is interesting for the present analysis is the role he attributes to speed and slowness in the process of the improvement of the quality of human life between the buildings and the correlation of his postulates with the "spatial" postulates of the advocates of slow living.

"Slow" urban design, postulated by both Gehl and the members of the Slow City movement, takes special care for those places that offer a chance for slowing down, which are important for the human encounters without which an urban space cannot aspire to be a living space. No meaningful interaction is possible when speed takes over. Gehl observes that people walk at an average speed of 5 kph and that the efficiency of the human sensory apparatus is correlated with

---

79   Gehl, *Life Between Buildings*, pp. 9–14.

this speed.⁸⁰ The physical arrangement of space may either promote or prevent visual and auditory contacts between people. High speed is perceived as one of the main factors that impede contact in public spaces, while low speed is recognized as a factor which enables it. Moreover, an increase in the speed of movement affects the ability to differentiate details and to process significant social information. The dimensions of elements of the environment that have been built for cars are conspicuously bigger than those designed to be seen not by drivers but by pedestrians. An ability to relate directly to the city, which can be identified as one of the most important elements in the creation of people-oriented urban space, is determined not only by a chance to slow down but also by an awareness of the importance of focusing on "[g]ood ground floor facades [which] are an important city feature."⁸¹ Taking care for what is within the human sight and human reach reduces distance and builds a relation between the city and its resident. The facades of the buildings "make the city interesting to walk through, interesting to look at, to touch and to stand next to."⁸² They start telling their stories only when a driver turns into a walker, stroller of flâneur. "Life takes place on foot,"⁸³ says Gehl, and this remark must entail serious reconsiderations of the character of the changes brought about by the domination of car culture.

Conclusions

An important element of a relation between man and place, which allows one to appreciate place's specificity and singularity, is a recognition of the role that speed and acceleration play in constituing the identities of both individuals and the places they inhabit. The revalorization of slowness in cultural discourse helps highlight "the importance of creating, and being able to access, 'slow places' in our everyday contexts."⁸⁴ Approached from the perspective of urban design, slowness can be seen as a vital aspect of what Pierre Sansot calls the "urbanism of delay" which "without impeding the free circulation of people and goods, takes seriously the need to live, and thus to remain behind, in the places to which we have the closest connection."⁸⁵

---

80 Gehl, *Life Between Buildings*, p. 63.
81 Gehl, Gemzøe, *Public Spaces ...*, p. 32.
82 Gehl, Gemzøe, *Public Spaces ...*, p. 32.
83 Gehl, *Life Between Buildings*, p. 72.
84 Parkins, Craig, *Slow Living*, p. 73.
85 Pierre Sansot, "On the Proper Use of Slownes," trans. Jeffrey T. Schnapp in *Speed Limits*, ed. Jeffrey T. Schnapp (Milan: Skira, 2009), p. 298.

The modes of thinking about the meanings of place in an age of deteritorialization have been recently marked by conceptual changes generated by an exhaustion caused by the autocratic domination of the logic of flow and mobility that has subsumed other forms of interaction with space, especially those that are not necessarily oriented towards the promotion of fast movement. There has been a cultural readiness to legitimize movements and practices which defend "slow spaces" and generate conditions in which the spaces of the slow may function *next to* the spaces of the fast. From the domestic spaces of slow homes to public spaces of slow and slowed-down cities, contemporary spaces have become the subject of negotiations that allow one to challenge the premises of the culture of speed and to seek the pleasures of a different kind of inhabitation of space.

Bringing the issues of slowness and their cultural ramifications to debates about spaces accentuates the democratic character of the idea of the city. In *New City Life* we read: "[t]he city is a place for everyone to meet. There are no admission requirements and you don't need a ticket: everyone has access. As important as it is for the city to be a meeting place for everyone, that can only happen if the city is open and democratic and inclusive of society's many different members."[86] Although not free from utopian overtones, this description of the city highlights the cultural and psychological diversity of urban dwellers, which in the context of the discourse of slowness may suggest a desire to integrate various speeds within urban space. If a postulate of further democratization of urban spaces is to be successfully enacted, then spaces of slowness such as squares, pedestrian and cycling lanes, car-free zones and public gardens and parks need to be included in urban planning, as it is only through such reclaimed spaces and territories that a new type of (slow) urban identity can successfully emerge.

---

86   Gehl, Gemzøe, Kirknæs, Søndergaard, *New City Life*, p. 86.

Chapter Six: Negotiating Mobility: On the Slow Move

> To tour, to stop, to drive slowly, to take the longer route,
> to emphasise process rather than destination, [...].
> John Urry, *Mobilities*

> The key to slow travel is a state of mind.
> That can be developed at home.
> Nicky Gardner, "A Manifesto for Slow Travel"

> To really know why San Francisco is not Paris, you must *sense* it.
> Joy Monice Malnar, Frank Vodvarka, *Sensory Design*

In 1974 Ivan Illich created a disturbing image of a passenger as a victim of intoxication with technological advance, blinded by what he accepts uncritically as a blessing of the comfort of modern travel:

> The habitual passenger cannot grasp the folly of traffic based overwhelmingly on transport. His inherited perceptions of space and time and of personal pace have been industrially deformed. He has lost the power to conceive of himself outside the passenger role. Addicted to being carried along, he has lost control over the physical, social, and psychic powers that reside in man's feet. The passenger has come to identify territory with *the untouchable landscape* [emphasis mine] through which he is rushed. He has become impotent to establish his domain, mark it with his imprint, and assert his sovereignty over it. He has lost confidence in his power to admit others into his presence and to share space consciously with them. He can no longer face the remote by himself. Left on his own, he feels immobile.[1]

Although slow travel, recognized as a distinctive cultural phenomenon and a movement with its specificity embraced by its own manifesto and a growing number of publications that popularize the idea, emerged some time after Illich wrote *Energy and Equity*, his understanding of the conditions of travel today and the rise of the category of the "habitual passenger" may be read from the perspective of the anxieties and concerns which rest at the core of slow tourism. When approached from the perspective of velocentrism with its focus on the sense of change generated by speed, the "untouchable landscape" tells the story of the spatial alienation of a contemporary passenger who has internalized a wide range of cultural practices which, as John Tomlinson puts it, close the gap

---

1   Ivan Illich, *Energy and Equity* (New York: Harper & Row, 1974), p. 25.

between a desire and its attainment, lead to annulment of the very experience of travel as well as the landscape that is travelled through, and thus create geographies of disruption and discontinuity.

The slow living revolution, which challenges the premises of fast life and brings a defamiliarization of various spaces of social and cultural life dominated and shaped by the "tyranny of the moment," has succeeded in affecting an apparently hostile environment of (fast) mobility and has led to the rise of what is now referred to as *slow* or *soft* mobility. Slow mobility is not an oxymoron but rather an expression of the sense of disillusionment with modernity and the lifestyles that it has been constantly producing and renewing. The rise of slow mobility reveals a yearning for the possibility of living in another world in which other values are seen as objects of desire. Lyn Thomas seems to be right when she says that "[c]ontemporary consumers might more than ever before be living in one world while dreaming of another [...]."[2] Slow travel, an increasingly popular form of slow mobility, is emblematic of the readiness to bridge the gap between the world we live in and the world we *would like to* inhabit.

In the age of supersonic planes and high-speed trains, the concept of slow travel not only sounds like a contradiction in terms, but also immediately transfers us, contemporary travellers, to a distant past when all travel *had to* be slow because no other pace of travelling existed. Travelling slowly is associated with impatience, impediments to movement, traffic jams, road rage, delays, and cancelled meetings, and produces a general sense of an irritating waste of time. When released, however, from this chain of negative connotations, slow travel reveals a capacity to become a site of negotiation of fast mobility as well as a platform for expression of one's (tourist) identity.

In this chapter I shall examine "A Manifesto for Slow Travel" and the guidebook *Slow London* to discuss the rhetoric which marks the contours of the identity of a slow traveller, and the representations of the slow tourist experience that these texts construct. I shall look at the ways in which they address and relate to some of the issues identified by recent studies in slow travel and tourism and pay special attention to the role of sensory experience in the act of slow travel. Finally, I shall discuss slow travel in the context of Kate Soper's theory of alternative hedonism, which opens up a space for questions of agency and resistance with relation to slow travel.

---

2   Lyn Thomas, "Ecochic: Green Echoes and Rural Retreats in Contemporary Lifestyle Magazines," in *The Politics and Pleasures of Consuming Differently*, eds. Kate Soper, Martin Ryle, Lyn Thomas (Basingstoke: Palgrave Macmillan, 2009), p. 72.

1. New Tourism

European modes of thinking about what travel is, what it should include and what should be eliminated from travel experience as unnecessary, burdensome, dangerous or simply uninteresting, have been changing in a manner which is illustrative of the evolution of our cultural consciousness. In the ancient world a voyage, whether literary or mythological, was the privilege of gods and heroes. The middle ages repeated the ambivalent character of travelling, which was seen either as something accessible only to the elite (going on a pilgrimage or crusade was an exception due to its availability to all social strata – medieval "tourists" were products of a Roman Catholic cultural tradition) or as a form of social activity which, when performed by social outcasts, had the potential to destabilize the social order of any community. Travel, in the broad sense of the word, could be afforded only by those who had either everything or nothing to lose. The political situation in medieval Europe also activated those who were pushed to travel by two powerful forces: military action and religious fanaticism. The crusades first, and geographical discoveries later, gradually extended the boundaries of the known world, but a journey remained an enterprise restricted to the few. Due to difficult travelling conditions, a journey was always time-consuming and dangerous, and in most cases required tolerance of discomfort. It called for physical, economic and psychological preparedness. Modernity has democratized travelling. One of the most significant characteristics of the process of democratization of travelling has been the rise of mass tourism, and in consequence a gradual reconceptualization of the meaning of travel as a kind of an obstacle on the way to the destination. Today a critical appreciation of the journey, which is stigmatized as time-consuming and inefficient, tends to be eliminated from tourist experience.

It is unsurprising that the Slow Food Manifesto, with its call to celebrate sensual pleasures, turned out to appeal to thousands of tired and disillusioned inhabitants of the fast new world, and for many the Slow Food movement became an alternative to mass-promoted models of life in this accelerated world. Formulated under conditions whose character has been shaped by the omnipresent imperative of instantaneous satisfaction of appetites, the idea of "long-lasting enjoyment" not only did not sound anachronistic, but seemed avant-garde, revolutionary and subversive. That the Manifesto grasped the essence of a general sense of disappointment with the results of cultural, social and political acceleration is evidenced not only by the movement's rapid growth in popularity, but also by the quick importation of its main ideas into other spheres of life.

Today, slow tourism and slow travel form, next to Cittaslow, one of the most visible "by-products" of the slow revolution that started in 1989. The concept of "gastronomic tourism" Carlo Petrini introduced and defined in terms of the readiness "to be aware and well-informed about the places visited: respectful, slow, reflective, and as distant as possible from the culture of 'use and discard'"[3] revealed its power to initiate changes. Transplanted and modified to meet and challenge the specificity of contemporary tourism, slow travel has developed to be a widely discussed social and cultural practice.

The evolution of slow travel is sometimes described as eclectic.[4] This claim points towards the diversified sources that inspired the phenomenon's rise and reflects the multidimensional character of the idea of travelling slowly in an age of high-speed trains and planes. It is commonly believed that slow travel has emerged under the influence of the Slow Food philosophy but to see it only as a by-product of Petrini's movement would be to fail to register various connections between slow travel and the increase in pro-environmental awareness, the travelling tradition and contestation of mainstream lifestyles.

The last two decades have witnessed a steady growth in the phenomenon, which aims at the rehabilitation of the significance of the physical act of travelling seen as a quality experience. Recognized as "an emergent market segment" at the 2007 London World Travel Market and forming 10% of the holiday market,[5] slow travel is more than a tourist trend. Its social and cultural ramifications encourage us to inscribe the phenomenon in a wider context of debates concerning the cultural dimensions of acceleration, globalization and ecology.[6] According to Euromonitor International it will soon become "a significant alternative to 'sun and sea' and cultural tourism,"[7] and it is reported that there will

---

3   Carlo Petrini, *Slow Food: The Case for Taste*, trans. William McCuaig (New York: Columbia University Press, 2001), p. 57.
4   Janet E. Dickinson, Les M. Lumsdon, Derek Robbins, "Slow Travel: Issues for Tourism and Climate Change," *Journal of Sustainable Tourism*, Vol. 19, No. 3 (2011), p. 283.
5   Les M. Lumsdon, Peter McGrath, "Developing a Conceptual Framework for Slow Travel: A Grounded Theory Approach," *Journal of Sustainable Tourism*, Vol. 19, No. 3 (2011), p. 265.
6   Lumsdon and McGrath's research shows that for the experts interviewed slow travel and slow tourism are two dimensions of the same phenomenon: the experts refer to "slow travel as the journey," while slow tourism is used to describe "a way of enjoying the destination. They are one and the same." Lumsdon, McGrath, "Developing a Conceptual Framework...," p. 274.
7   Lumsdon, McGrath, "Developing a Conceptual Framework...," p. 266.

have been an estimated annual 10% increase in the growth of slow tourism by 2015.[8]

It is generally agreed that four general strands can be distinguished to embrace the main issues of slow travel: an escape from the tyranny of time regime, an emphasis on the destination in terms of the search for a sense of place and locality, revalorization of the experience of the journey, and a concern with ecology informed by a desire to "travel without thrashing the planet," to quote Greentravelguides.tv ("the world's first television channel dedicated to responsible tourism,"[9] as they describe themselves on their website). The "eclectic evolution" of the phenomenon, as well as the fact that slow travel is far less defined by regulations, laws and organizational and administrative structures than any other form of organized slow living, accounts for the fact that it still inspires attempts to formulate not so much a definition but rather "a group of associated ideas."[10] Janet Dickinson defines slow travel as "an emerging conceptual framework which offers an alternative to air and car travel, where people travel to destinations more slowly overland, stay longer and travel less."[11] A more elaborate definition is provided by Les Lumsdon and Peter McGrath, who say that

> [s]low travel is a sociocultural phenomenon, focusing on holidaymaking but also on day leisure visits, where use of personal time is appreciated differently. Slowness is valued, and the journey is integral to the whole experience. The mode of transport and the activities undertaken at a destination enhance the richness of the experience through slowness. Whilst the journey is the thing and can be the destination in its own right, the experience of locality counts for much, as does reduced duration or distance of travel.[12]

"[I]nterpretations of slow travel are diverse,"[13] says Dickinson et al. and this fact, illustrative of the phenomenon's eclecticism, provides a point of departure for further theorization of slow travel. Exhaustive as they seem to be, definitions of slow travel encourage one to ask further questions concerning, among other things, the modes of enhancement of the richness of the slow tourist experience as well as its epistemological and ontological dimensions, and the ways in which the identity of a slow tourist is constructed by slow travel texts and by slow

---

8   Lumsdon, McGrath, "Developing a Conceptual Framework…," p. 266.
9   *Green Travel Guides TV*, accessed 8 February 2014, http://www.greentravelguides.tv/
10   Lumsdon, McGrath, "Developing a Conceptual Framework…," p. 273.
11   Janet Dickinson, quoted by Janet Dickinson, Les Lumsdon in *Slow Travel and Slow Tourism* (London: Earthscan, 2010), p. 1.
12   Lumsdon, McGrath, "Developing a Conceptual Framework…," p. 276.
13   Dickinson, Lumsdon, Robbins, "Slow Travel…," p. 282.

tourism practices. The focus in this chapter is not on evaluation of the extent to which slow travel brings a change to greenhouse gas emission (though slow travel advocates have no doubt that it leads to the reduction of the carbon footprint), but on the ways in which slow travel is used in the process of self-definition. Dickinson rightly observes that "[t]ourism is one of the mechanisms through which people are able to present their identity to others,"[14] and emphasizes the role slow travel plays in the process of negotiation of one's (tourist) identity in the age of mass mobility and mass tourism and the ways in which it becomes an important platform for self-identification.

2. The Return of the Repressed: Travelling in the Travelless Age

Analyzed in the framework of the history of tourist experience the present cultural moment can be defined in terms of transition from travel*free* to travel*less* age. I propose to use the term "travelfree" for it points at the sense of celebration of the liberation of an individual from the burden of travel. Seen on one hand as hard work (the word "travel" comes from Old French *travailler* and means to work hard) and on the other hand as a waste of time in the culture of rush, travel has been represented as an unnecessarily burdensome practice, or, as Nicky Gardner puts it in "A Manifesto for Slow Travel," "a minor inconvenience"[15] whose elimination has been cast as a marker of comfort and maximization of efficiency, a value central to modern axiological reality. The sense of comfort and pleasure in the travelfree age originates, largely, in an act of annulment of the experience of travel.

The transition from travelfree to travelless age is marked by the rise of the phenomenon of slow travel. Illustrative of the sense of yearning for the experience of travel, which has been systematically and successfully erased from the spectrum of contemporary experience, slow travel tells a story which is underwritten by a sense of deprivation and lack, the story of a quest for something that is believed to have been lost. Seen in this light, slow travel can be approached as a form of alternative hedonism with its roots in the sense of displeasure and dissatisfaction with the models of life promoted by hyper-consumerism. As Kate Soper has pointed out: "the affluent lifestyle is generating its own specific forms of disaffection, either because of its negative by-products or because it stands in

---

14   Dickinson, Lumsdon, Robbins, "Slow Travel…," p. 295.
15   Nicky Gardner, "A Manifesto for Slow Travel," *Hidden Europe*, No. 25 March/April (2009), p. 10. Throughout the chapter Gardner's manifesto will be referred to as *AMFST*.

the way of other enjoyments."[16] The description of the character of the present cultural moment in terms of its "travellessness" registers a meaningful change in the conceptualization of travelling experience and allows us to address issues as diverse as agency, environmental consciousness and "other pleasures" as forms of contestation of acceleration and the "supposed blessings of consumerism."[17]

"A Manifesto for Slow Travel" by Nicky Gardner, the editor of *Hidden Europe* magazine, is one of the most explicit representations of slow travel ideas. Regardless of the character of the cause – be it social, political, cultural or artistic – a manifesto[18] is always an expression of a certain ontological certainty which is constructed by means of accentuating a difference. In her analysis of the Slow Food Manifesto, Wendy Parkins refers to Janet Lyon's *Manifestoes: Provocations of the Modern* whose insightful reading of the nature of manifestoes provides her with the theoretical framework, to emphasize that manifestoes are "the means by which movements articulated their own identity by rhetorical denunciation of their enemies."[19] Lyon demonstrates that a manifesto has always been a forum for the expression of the needs of those who felt they had been marginalized, oppressed and enslaved: "to write a manifesto is to announce one's participation, however discursive, in a history of struggle against oppressive forces."[20]

Two main components of a manifesto which play a significant role in the process of the creation of a certain vision of the world include a representation of the world/community/movement, which is seen as desirable or beneficial for a group, community, or the whole of mankind, and a counter-representation which involves a critical representation of the *status quo*. Each manifesto stems from the sense of dissatisfaction with the present (or some aspects of it) and presents a "method" by means of which the elimination of negative aspects of the present is not only deemed possible but also strongly recommended.

---

16 Kate Soper, "Alternative Hedonism, Cultural Theory and the Role of Aesthetic Revisioning," *Cultural Studies*, Vol. 22, No. 5 (2008), p. 571.
17 Soper, "Alternative Hedonism …," p. 571.
18 In this chapter I refer to Janet Lyon's analysis of the character of manifestoes in *Manifestoes: Provocations of the Modern* (New York: Cornell University Press, 1999). The analysis of "A Manifesto for Slow Travel" from the perspective of Lyon's study of manifestoes as an ideologically framed genre owes its general shape to Wendy Parkin's reading of the Slow Food Manifesto in "Out of Time: Fast Subjects and Slow Living," *Time & Society*, Vol. 13 No. 2/3 (2004).
19 Wendy Parkins, "Out of Time: Fast Subjects and Slow Living," *Time & Society*, Vol. 13 No. 2/3 (2004). p. 372.
20 Janet Lyon, *Manifestoes: Provocations of the Modern* (New York: Cornell University Press, 1999), p. 10.

The character of a manifesto demands an appropriate rhetoric: the either/or logic which permeates a manifesto as a central rhetorical device is determined by the aim it is to achieve. The presence of "rigid hierarchical binaries,"[21] which is a defining trait of a manifesto, makes it clear that it is not a means of negotiation. It is a tool of resistance and persuasion and as such must rely on sharp contrasts and appealing and attractive accentuation of difference. Lyon refers to this in terms of "rhetorical straightforwardness"[22] which in the case of "A Manifesto for Slow Travel" is reinforced by a clearly-marked demarcation line which separates "us" from "them." The contrast between mass tourism and slow tourism is accentuated by the construction of images and metaphors representing slow travel as the manifestation of one's individuality and ability to liberate oneself from the power of cultural clichés. These, in turn, are believed to determine the character of tourist needs and help form the list of tourist attractions and priorities, which is later methodically ticked off in what Gardner calls "a pilgrimage of mass consumption" (*AMFST*,12). Such a polarization of tourist experience is given symbolic representation. The iconography of slow travel oscillates between two major images: a donkey, "an indispensable asset of the would-be slow traveller," and planes, those "fragile aluminium tubes which […] shoot through the sky at slightly less than the speed of sound" (*AMFST*,11). Both planes and high speed trains are cast as a slow traveller's enemies for the effect they have on the mind and body of a person on the move. The speed at which they move disconnects a traveller from the physical space he or she is travelling through and leads to the disappearance of the landscape, which becomes "untouchable" and as such is no longer seen as an indispensable condition for the travelling experience. A description of the experience of travelling by high-speed train provided by Gardner in her manifesto abounds in expressions that reflect states of disintegration, disconnection and fragmentation: the trains "slice" the space, "defying the warp and weft of the landscape" and produce "distracted passengers" who are exposed to "flashes of light between tunnels, angled glimpses of the sky and plenty of scope for headaches" (*AMFST*, 12). In contrast to the unnerving experience of fast travel, any journey by a slow train is represented as a source of radically different emotions as "the train follows the meandering course of the river, affording wonderful views of gabled villages, precipitous vineyards and romantic gorges" (*AMFST*, 12).

According to Gardner, slow travel is "about having the courage not to go the way of the crowd" (*AMFST*, 12) and while mass tourism is represented as

---

21   Lyon, *Manifestoes…*, p. 3.
22   Lyon, *Manifestoes…*, p. 2.

consisting in the unending repetition of tourists schemes, whose stability is guaranteed by the creation of tourists needs that the schemes are supposed to satisfy, "the slow traveller does not play the same dangerous game" (*AMFST*, 12). The opposition thereby erected positions a slow traveller as a cultural rebel and anarchist, not likely to be manoeuvred into the traps set up by a culture which erases any traces of individual taste and produces a homogenized mass of tourists doomed to participate in a spectacle of "numbing boredom" (*AMFST*, 12). The appropriation of the either/or opposition as well as Gardner's manifesto's "rhetorical mode of directness, its deployment of a declarative, passionate voice,"[23] allows for the creation of the ideological framework within which to place slow travel. The aesthetic and cognitive "unattractiveness" of fast travel is contrasted with the valorization of slow travel, which is represented as being capable of providing the foundations necessary not only for the rehabilitation of the meaning of travel but also for the construction of a model of contestation of mass culture, its products, lifestyles and ideologies.

Gardner says: "travel has somehow slipped out of fashion. […] The pleasure of the journey is eclipsed by anticipation of arrival. To get there fast is better than to travel slow" (*AMFST*, 10). Devaluation of travel is seen as symptomatic of the acceleration of everyday life. Gardner has no doubt that "[m]odernity comes at a cost" (*AMFST*, 11) and these words acquire a new dimension when we read them in the context of Lyon's analysis of a manifesto as "the form that exposes the broken promises of modernity."[24] It is not too much to say that one of the promises of the technologically advanced culture of modernity was the comfort of everyday life, which consists, among other things, in the acceleration of what used to be slow. The point is that "the acceleration of just about everything,"[25] to quote the title of James Gleick's study, this dream that came true, has proved to be a mixed blessing.

Travelling denotes a disturbing state of "in-betweenness" which contemporary culture systematically tries to annihilate. Thomas Mann's *The Magic Mountain* provides an unexpected insight into the very heart of the anxiety caused by enforced suspension of the rhythm of fast, industrial reality. The novel opens with a description of the journey to Davos made by Hans Castorp, a young German engineer. The scene, reflexive and apparently un-dynamic in spite of the movement of the train, becomes a pretext to contemplate the nature of the

---

23   Lyon, *Manifestoes…*, p. 14.
24   Lyon, *Manifestoes…*, p. 3.
25   James Gleick, *Faster: The Acceleration of Just About Everything* (New York: Pantheon Books, 1999).

experience of space, which in the beginning of the 20th century started very energetically to make its way into the consciousness of modern man. Mann begins:

> An ordinary young man was on his way from his hometown of Hamburg to Davos-Platz in the canton of Graubünden. It was the height of summer and he planned to stay for three weeks.
>
> It is a long trip, however, from Hamburg to these elevations – too long, really, for so short a visit. The journey leads through many a landscape, uphill and down, descends from the high plain of southern Germany to the shores of Swabia's sea, and proceeds by boat across its skipping waves, passing over abysses once though unfathomable. [...]
>
> Two days of travel separate this young man (and young he is, with few firm roots in his life) from his everyday world, especially from what he called his duties, interests, worries, and prospects – separate him far more than he had dreamed possible as he rode to the station in a hansom cab. Space, as it rolls and tumbles away, between him and his native soil, proves to have powers normally ascribed only to time; from hour to hour, space brings about changes very like those time produces, yet surpassing them in many ways. Space, like time, gives birth to forgetfulness, but does so by removing an individual from all relationships and placing him in a free and pristine state – indeed, in but a moment it can turn a pedant and philistine into something like a vagabond. Time, they say, is water from the river Lethe, but alien air is a similar drink; and if its effects are less profound, it works all the more quickly.[26]

When in 1907 Castorp travelled from Hamburg to Davos it took him two days to cover a distance of about 750 kilometers. Today the same distance can be covered in ten hours by train or car, and one hour and thirty minutes by plane. When approached from the perspective of travelling experience, Mann's description of Castorp's journey strikes one not only as an interesting illustration of the change Western culture underwent in a relatively short period of time but also as an illustration, paradoxically, not so much of the phenomenon labeled "slow travel," though at the first sight it may appear to be so, but rather of the sense of unease associated with not being fast enough.

Hans Castorp is a liminal figure who belongs to two worlds at the same time. As an engineer and materialist, he is aware of the meaning of time for a man of work, a producer whose efficiency has a measurable value. He knows that time is money and a two-day journey to the Alpine sanatorium, where he plans to spend three weeks only, is to some extent irrational, being "too long, really, for so short a visit." His perception of the meaning of the duration of the travel is deeply

---

26 Thomas Mann, *The Magic Mountain*, trans. John E. Woods (New York: Vintage, 1996), pp. 3–4.

rooted in the ideology of maximization of profits. On the other hand, Hans Castorp is a man whose contemplative nature makes him see the world as a site of experiences whose nature far exceeds the logic of gains and losses. The journey to Davos is an experience in itself; it is permeated by the sense of space and time and the influence they have on an individual. His two-day train journey reveals the power of the influence they exert. Space, as experienced by Castorp, does not undergo annihilation and annulment; on the contrary, it begins to exert its influence on him.

The slow covering of the distance is a factor which situates a traveller in the space *between*: between the place a traveller has just left and his destination. An accelerated culture with its focus on instant gratification of man's every need deems such a state of "in-betweenness" to be worthless and thus in need of immediate shortening. The two-day journey to Davos reveals its uprooting potential and turns out to be counterproductive. The "forgetfulness" that Mann writes about, which is the effect of the process of uprooting generated by a long journey, is a dangerous luxury which has a transformative function: it changes a pedant and philistine into a vagabond. The journey suspends the traveller and the principles of his world; it suspends the legitimacy and applicability of the principles, which form the foundation of the sense of order, and coherence of the world the traveller has been separated from. The anarchy of slow travel reaches its climax at the moment a vagabond is born; by escaping the rigid frames of the social order he situates himself "outside of the obligations of place and roots,"[27] thus becoming a source of social instability and disruption.

Castorp's anxiety foreshadows present-day dedication to the cause of shortening the time spent waiting, which today is associated with waste of money and inefficiency. Slow travel is a challenge to task-oriented mass tourism, whose appeal includes maximization of tourist experience, which often produces a sense of accomplishment realized by seeing everything that *has to* be seen. Slow travel and slow tourism move against this flow with their celebration of "the other pleasures" and challenge the authoritarian rule of the consumerist model of the good life by prioritizing models of non-commodified gratification and emphasizing the ideas of exchange, reciprocity and co-responsibility and the role they play in marking the contours of post-consumerist pleasure. The rhetoric of slow travel goes against the flow of market-oriented tourist practices by encouraging the rise of structures of feeling with roots in the growing sense of

---

27 Tim Creswell, *On the Move. Mobility in the Modern Western World* (New York: Routledge, 2006), p. 11.

displeasure generated by the side-effects of the domination of the models of life promoted by present-day culture. Slow travel situates itself in a new, hedonist imaginary whose aim is to "subvert current perceptions of the attractions of a consumer material culture."[28] Approached from the perspective of an "'alternative hedonist' disenchantment with consumerism,"[29] slow travel becomes emblematic of a concept of sustainable life which is marked by a decisive departure from high consumption and the immediate satisfaction of needs. As an act of alternative hedonism it aims to "accommodate the goods that are currently being lost or marginalised."[30] Soper believes that the quest for an alternatively defined pleasure should involve "very different conceptions of consumption and human welfare from those promoted under capitalism."[31] Gardner's description of slow travel with its stress on a redefinition of the nature of basic elements of the tourist experience is informed by a similar attitude:

> Slow travel is about making conscious choices. It is about deceleration rather than speed. The journey becomes a moment to relax, rather than a stressful interlude imposed between home and destination. Slow travel re-engineers time, transforming it into a commodity of abundance rather than scarcity. And slow travel also reshapes our relationship with places, encouraging and allowing us to engage more intimately with the communities through which we travel (*AMFST*, 11).

The advocates of the idea of slow travel see it as a "deeper form of a travel" which is not only slower but also easier and simpler.[32] Two main suggestions of slow travelling include spending at least one week in one place, preferably in a vacation rental and not in a hotel, and using the "Concentric Circles" theory for day trips.[33] This model of slow leisure, integrally linked with slow travel, assumes liberation from the pressure of the model of standard tourism, which is task-and-quantity oriented and whose framework is defined by "must-sees" which often entail racing through the trip. Slow travel replaces the standard model of European leisure, which is seen not only as most popular but also most "natural," with quality-time oriented models of slow leisure. What turns out to be one of the most characteristic features of the phenomenon is that slow travel is often a

---

28 Soper, "Alternative Hedonism...," p. 567.
29 Soper, "Alternative Hedonism...," p. 567.
30 Soper, "Alternative Hedonism...," p. 572.
31 Kate Soper, *What is Nature? Culture, Politics and the Non-Human* (Oxford: Blackwell, 2000), p. 271.
32 "What is Slow Travel?" accessed 10 February 2014, http://www.slowtrav.com/vr/index.htm
33 "What is Slow Travel?"

natural extension of lifestyles practiced at home. It is not so much a two-week experiment carried out during leisure time but rather a consequence of lifestyle choices made with reference to ethically oriented definitions of sustainable life in the context of hyper-consumerism. More often than not, it is a manifestation of a certain mode of thinking about the world and the role one plays in the world's system of interdependency.

Slow travel philosophy makes a distinction between "staying" and "living" in a given destination a fundamental principle. "Living" in a place assumes that a traveller has time and willingness to discover the specificity of the place and is ready to become a part of the locality for a short period. "Living" assumes connection, while the idea of "staying" as seen by slow travel enthusiasts is associated with disconnection, superficiality and passing through. Living in a place assumes integration into the cityscape rather then being "merely a passing observer" (*AMFST*, 12).

Slow travel is cast as an exercise in mindfulness and an act of liberation from the oppressive and confining understanding of time developing over the last few hundred years. It signals a departure from mass tourism and the homogeneity it offers. At the basis of slow travel lies a redefined attitude towards time, which is no longer seen as a scarce commodity to be obsessively monitored and used economically. For a slow traveller time spent travelling from place to place is never lost and it is just as important as the time spent in one's destination. Getting slowly there is as meaningful an experience as being there. Slow travel allows one to rediscover the meaning of travel, which is no longer seen as "a stressful interlude imposed between home and destination" but as a "moment to relax" which "re-engineers time, transforming it into a commodity of abundance rather than scarcity" (*AMFST*, 12). No enemy of timetables, clocks and timely arrivals, slow travel reclaims and liberates time from their tyranny.

Slow travel in the age of acceleration is about a connection – to a place, to its people and to culture[34] – and awareness of mutual dependency is one of the ideological ramifications of slow travel. A dedication to the cause of the welfare of the planet and its inhabitants dismantles the boundaries between "us" and "them." Those committed to the idea of consuming differently do not perceive themselves as "innocent and passive victim[s] of industrialism;" they are motivated by the belief in the power of the choice when

> as consumers they opt wherever possible for fair trade and more environmentally-friendly goods or services, to cut down on packaging and plastic use wherever possible,

---

34   "What is Slow Travel?"

to spend time cooking rather than use fast food, to avoid whenever possible the more ecologically damaging forms of holiday and leisure activity, and so on.[35]

Environmental consciousness forms an essential dimension of slow mobility, which can be seen as a form of the "green pleasure" that has been defined by Richard Kerridge as the type of pleasure that "follows the logic of environmentalism – by using less carbon, deepening one's love of things already at hand, appreciating cycles of growth and renewal in the local and global ecosystems, understanding and taking delight in interdependency."[36] Soper's alternative hedonism and Kerridge's green pleasures are informed by a sense of reciprocity. The quest for a redefined pleasure is embedded in a desire to see oneself in the broader context yet still make space for one's individual concerns. Slow travel is held to be the source of "other pleasures" which flow from a sense of co-responsibility for the place and locality and from a readiness to engage with the local, not only by accepting but also by giving and sharing, which often take the form of ecologically-oriented modes of behavior as well as support for the local economy and agriculture.

3. Smelling the Roses: From the Occularcentric to the Multisensory

In "A Manifesto for Slow Travel" we read:

> It is easy to practice slow travel. Start at home. Explore your immediate locality. Leave the car at home and take a local bus to a village or suburb that perhaps you have never visited. Plan expeditions that probe the jungles and traverse the deserts of your home neighbourhood. Visit a church, a community centre, a cafe, a library or a cinema that you have passed a thousand times but never entered (*AMFST*, 13).

Deceleration can create "slow spaces" where no one would expect them to flourish and proves that the "slow metropolis" is not a destination impossible. Slowness, this modern catch-phrase, enters more and more daringly new spaces and conquers new territories, thus making the idea of "slow London" sound less like a ridiculous oxymoron than a trendy phrase which reflects London's ability to adopt the latest trends and cultural practices. Harry Eyres, the author of the Slow Lane column for the *Financial Times*, believes that slowing down in London

---

35 Soper, "Alternative Hedonism…," p. 572.
36 Richard Kerridge, "Green Pleasures," in *The Politics and Pleasures of Consuming Differently*, eds. Kate Soper, Martin Ryle, Lyn Thomas (Basingstoke: Palgrave Macmillan, 2009), p. 131.

"encourages us to raise our heads from contemplation of the paving stones. Just a small change in the angle of our view can bring unexpected delights into focus. London's ancient fabric is shot through with crevices where slowness can settle."[37] In the words of Carl Honoré, "London is the perfect place to slow down" because "[i]t combines the volcanic energy of a big city with architecture, green spaces and villagey streets that encourage us to stop and stare."[38] The London which the authors of the project re-discover, re-claim and wish to popularize resembles the city that appears in the blockbuster movie *Notting Hill*, a place where life is slower, streets are narrower and interpersonal contacts natural, regular, spontaneous and always carried out at the table.

Such cultural events as Slow Down London Festival and Slow Down London Day give Londoners a new form of public space activity and educate through entertainment. The education offered on such occasions challenges the main premises of fast life according to which there is no retreat from the world of acceleration and speed is the guarantee of efficiency. Instead, it invites one to reconceptualize the role of speed by embracing various sources of slowness, from the much-celebrated and world-famous slow food, through slow art and meditation to slow travels and crafts. The rehabilitation of slowness consists in the establishment of the connection between slowness and improvement.

It comes as no surprise that in the wake of the recent explosion of interest in anything slow, slow guidebooks have started to appear and reveal a growing appetite for slowness in tourism. Published in 2010 by Affirm Press and Hardie Grant Books *Slow London*, described by the publishers as "an inspirational lifestyle guide for Londoners who want to live more and fret less,"[39] is one in a series of guidebooks which "are for anybody who wants to slow down and live it up. They celebrate all that's local, natural, traditional, sensory and most of all gratifying about living in each of these corners of the world."[40] All the books published in the series have a similar design featuring a man on a donkey, the donkey being another symbol – along with the snail – enthusiastically adopted worldwide by

---

37 Harry Eyers, "London's First Slow Down Festival," *Financial Times*, 28 March, 2009, accessed 10 February 2014, http://www.ft.com/intl/cms/s/0/b70c963a-1a5d-11de-9f91-0000779fd2ac.html#axzz2svJMTTvx
38 *Slow Planet*, Archives for December 2008, "Slow Down London," accessed 10 February 2014, http://www.slowplanet.com/blog/2008/12/page/3/
39 Robin Barton, Hayley Cull, *Slow London* (Melbourne: Affirm Press/Hardie Grant Books, 2010). Throughout the chapter this book will be referred to as *SL*.
40 Affirm Press, "Slow Guides," accessed 10 February 2014, http://www.affirmpress.com.au/slow-guides

the promoters of slow living, below which appears the slogan: "[l]ive more, fret less." The underlying suggestion of the project is that the integration of slowness and tourism is not so much about physical movement to far-away and exotic places, but rather creates a form of inner journey, the pre-requisite for which is slowing down.

In *44 Letters from the Liquid Modern World* Zygmunt Bauman refers to Walter Benjamin's differentiation between two types of stories: sailors' stories and peasants' stories.[41] While the former focus on the exotic, unimaginable, unheard of, distant and improbable, the latter revolve around the everyday, the usual, the close and "apparently familiar." Bauman writes: "I said '*apparently* [emphasis original] familiar' since the impression of knowing such things thoroughly, inside out, and therefore expecting there to be nothing new to be learnt from and about them, is also an illusion – in this case coming precisely from their being too close to the eye to see them clearly for what they are."[42] Bauman's intention is to narrate the world and contemporary experience from the perspective of "*sailors*' stories as told by *peasants* [emphasis original]."[43]

The story in *Slow London* is a sailor's story told by a peasant, which starts at home, and, just as Gardner recommends, at home it ends. As a guide to London it is untypical for it is intended to be read firstly by those who actually *live* in the city (although the guide can be used by all non-Londoners who are interested in discovering the "slow metropolis"). Addressing Londoners and making them the book's target audience is indicative of the assumption behind a decision to write a guide to the city for the city's inhabitants: it redefines them as strangers to the city and the city itself as *terra incognita*, a place that needs to be rediscovered and renamed and whose invisible slow side is there for those who know where to look for it. *Slow London* redraws the cartography of the city and by so doing deprives it of the sense of familiarity which tends to draw an invisible curtain over the city since, as Bauman says, "[n]othing escapes scrutiny so nimbly, resolutely and stubbornly as 'things at hand,' things 'always there,' 'never changing.' […] Their ordinariness is a blind, discouraging all scrutiny."[44] When not taken for granted and approached in a "slow adventure" mood, slow London displays its uniqueness and teaches one that "[a]fter all, even in the city of eight million people, it's possible to slow down and smell the roses" (*SL*, 7).

---

41   Zygmunt Bauman, *44 Letters from the Liquid Modern World* (Cambridge: Polity Press, 2010), p. 3.
42   Bauman, *44 Letters…*, p. 3.
43   Bauman, *44 Letters…*, p. 3.
44   Bauman, *44 Letters…*, p. 3.

Already referred to in Chapter Four, Slow Food's endorsement of the "primacy of sensory experience," which allows it to treat "eyesight, hearing, smell, touch, and taste as so many instruments of discernment, self-defence and pleasure,"[45] turned out to be constitutive of slow travel's epistemological foundations. The *Slow London* guide follows the main assumptions of Carlo Petrini's project of the rehabilitation of the senses by splitting the second part of the book, entitled "Be – slow down and smell the roses," into five sections, entitled respectively "See," "Hear," Smell," "Taste" and "Touch," where the five senses are represented as the tools of embodied perception. Slow guidebooks challenge the hegemony of sight which was long ago established in the Western sensorium. The guide's shift towards the wide spectrum of sensory experience of the urban space can be seen as an element of the new sensitivity to urban space whose essence is focused on the encouragement of urban designers to "disregard their occularcentric perspectives and reconceptualize the city as a multisensory space."[46]

Slow guidebooks defamiliarize well-known spaces and seek to engage the senses in the act of tourist exploration, which can no longer be narrowly described in terms of "sightseeing," since what is now being promoted should rather be called "bodyexperiencing." The engagement of all the senses in the practice of tourism broadens the formula of "getting to know" a place and stresses the corporeal aspect of any form of travelling. Traditional "fast" guidebooks favor a tourist scheme in which the purpose of going to a place is an imperative to see as much as possible, while slow guidebooks stress the qualitative (and sensory) dimension of tourism and replace popular "visiting" with "experiencing" a place.

The turn towards sensory experience opens up new cognitive spaces and creates new exploratory possibilities. To illustrate the way the senses can be reactivated through slow tourism, let us take a look at two sections of the guidebook entitled respectively "Hear" and "Touch." The sense of hearing engaged in the process of getting to know London rediscovers a variety of "soundscapes" and makes it possible to create an "aural archive of London" (*SL*, 83). Ian Rawes, the founder of the London Sound Survey, registers the diversity of the sounds of London which include, among other things, the sound of traffic, the sound of people but also a bird's song, and reveals the potential of the sound to tell the story of continuity and change. In the section entitled "Rhythms, Riffs and Recording" Hayley Cull writes that

---

45 Carlo Petrini, *Slow Food: The Case for Taste* (New York: Columbia University Press, 2001), p. 69.
46 Sarah Pink, "Sensing Cittáslow: Slow Living and The Constitution of the Sensory City," *Senses & Sociology*, Vol. 2, Issue 1 (2007), p. 65.

sounds can represent fashion, from singing canaries and wind chimes to boom-boxes and car horns that play the first eight notes of 'Colonel Bogey.' They reflect developments in trade, industry and technology; growth of the city itself; demographic and social change; even shifts in the scattering of wildlife, like the raspy squawk of the ring-necked parakeet as it becomes astonishingly prevalent here (*SL*, 83).

The section of the book called "Schedule of sounds" creates a calendar of "aural events" which includes "cascading chimes of Big Ben" on January 1st, "the lazy drone of bumblebees" in May and fireworks in November (*SL*, 85–89). The sonic aspects of the city life included in the guidebook correspond with David Pocock's study in which he says that "sounds play a crucial role in the anticipation, experience and remembering of places"[47] and emphasizes that they are often engaged in the acts of endowing places with meanings and values.

The recreation of the city's soundscape is represented as a source of uncommodified pleasure, one that allows for the rediscovery of the ability to hear, which paradoxically has been lost in the culture of noise. Similarly, touch, "the first sense we experience and the last we lose,"[48] undergoes rehabilitation as an important tool of cognition which, when revalued, opens the door to "the other London" and invites one to connect with time and tradition as "through marks of touch we shake the hands of countless generations."[49] Inviting us to touch our way through London, the authors of *Slow London* say:

> A smooth, polished feel dominates the ancient hardwood pews of St Paul's, while the battered timber bars of our old pubs are equally sacred. [...] Running your hand over the beautiful vines today is to connect with both a time and a tradition.
>
> Metal can feel colder but it's an equally good conductor of memories and stories. In places like the National Gallery and the Royal Courts of Justice, the banisters and handrails have been burnished by countless palms. Running soft hands along the worn railings is a timely illustration of how billions of tiny actions can have an effect on something as hard and intractable as metal or stone [...] (*SL*, 131).

The above excerpt suggests an interesting parallel with the "Preface" of the October 1991 issue of *The Architectural Review*, in which we read:

> We appreciate a place not just by its impact on our visual cortex but by the way in which it sounds, feels and smells. Some of these sensual experiences elide, for instance our full understanding of wood is often achieved by a perception of its smell, its texture (which

---

47   David Pocock, quoted by Joy Monice Malnar, Frank Vodvarka in *Sensory Design* (Minneapolis: University of Minnesota Press, 2004), p. 140.
48   Malnar, Vodvarka, *Sensory Design*, p. 127.
49   Juhani Palasmaa, quoted by Malnar, Vodvarka in *Sensory Design*, p. 145.

can be appreciated by both looking and feeling) and by the way in which it modulates the acoustics of the space.[50]

What both excerpts have in common is the revalorization of sensory experience underwritten by the conviction verbalized by Yi-Fu Tuan that "[t]aste, smell, and touch are capable of exquisite refinement. They discriminate among the wealth of sensations and articulate gustatory, olfactory and textural worlds."[51] Both *Slow London* and *The Architectural Review* invigorate the sensuous discourse by recasting senses as valid "ways to knowledge."[52] The senses have always been objects of skepticism and ambivalence and Western though has been informed by an oscillation between "the superiority or inferiority of the senses with respect to reason, and the ratio or balance of the sensorium."[53] Sarah Pink notes that it is emblematic of contemporary shifts in interest and the extension of research areas in social and cultural anthropology as well as cultural and literary studies that there is an increasing sensitivity to the fact that "the way people experience their environment is inevitably multisensory."[54] Rehabilitation of the sensuous experience in many spheres of life "has made it clear that sensory experience of a particular type of physical and material environment […] is inextricable from the cultural knowledge and everyday practice through which localities are constructed and experienced."[55]

What *Slow London* proposes is cognition on an intimate scale based on sensory perception. While Malnar and Vodvarka claim that "we need to revalue the nonvisual senses and learn a new vocabulary as well,"[56] Juhani Pallasmaa observes that "the city of the gaze passivates the body and the other senses, the alienation of the body again reinforces visibility."[57] *Slow London* reimagines the capital as a sensory city – the sense of pleasure is achieved by enhanced sense-oriented experience and overturning of the hegemony of vision. "The primacy of sensory experience" that Petrini talks about can be seen as one of the defining features of slow tourist experience, which thus can be understood

---

50   Malnar, Vodvarka, *Sensory Design*, pp. 23–24.
51   Yi-Fu Tuan, *Space and Place: The Perspective of Experience* (Mineapolis: University of Minnesota Press, 1977), p. 10.
52   Anthony Synnott, *The Body Social. Symbolism, Self and Society* (London: Routledge, 1993), p. 128.
53   Synnott, *The Body Social*, p. 128.
54   Pink, "Sensing Cittáslow…," p. 61.
55   Pink, "Sensing Cittáslow…," p. 61.
56   Malnar, Vodvarka, *Sensory Design*, p. 239.
57   Juhani Pallasmaa, quoted by Pink in "Sensing Cittáslow…," p. 60.

as "multisensorial and as such *neither* dominated by *nor* reducible to a visual mode of understanding [...]."[58] In the words of Yi-Fu Tuan, "[a]n object or place achieves concrete reality when our experience of it is total, that is through all the senses as well as with the active and reflective mind."[59]

The turn towards the sensory experience of slow travel corresponds with Soper's analysis of alternative hedonism when she says that it "does not reside exclusively in the desire to avoid or limit the un-pleasurable by-products of collective affluence, but also in the sensual pleasure of consuming differently."[60] *Slow London* can be seen as advocating a form of the experiencing of the metropolis which is based on the engagement of all the senses, thus giving rise to the "alternative urban 'sensescapes' that implicitly critique the visual, olfactory, gustatory, sonic and haptic experiences that are associated with global consumer capitalism."[61] The awakening of the redefined pleasure and the rise of alternative hedonism are deeply rooted in the realm of sensory experience.

The cognitive perspective adopted and promoted by the authors of *Slow London* correlates interestingly with a phenomenological view, the validity of the parallel supported by an explicit claim made by Gardner in her manifesto when she describes travelling as a "phenomenological process" and claims that "[s]low travel reinvigorates our habits of perception (*AMFST*, 13)." The city, Gardner writes, "deserves more than a casual glance – cityscapes are there to be studied and observed in detail. They are spurs to meditation and only much later can words flow" (*AMFST*, 13). The phenomenological character of the slow travel experience consists in the rehabilitation of the role the senses play in the process of learning. Considering slow travel's concern with ecological issues, it seems only natural that sensory perception should embrace "the perception of the environment," to refer to Tim Ingold's work, thus giving rise to ecophenomenological readings of the position and role of a human being in the world. David Abram, the author of *The Spell of the Sensuous*, believes that

> [o]ur bodies have formed themselves in delicate reciprocity with the manifold *textures, sounds and shapes* [emphasis mine] of an animate earth – our eyes have evolved in subtle interaction with *other* eyes [emphasis original], as our ears are attuned by their very structure to the howling of the wolves and the honking of geese. To shut ourselves off from these voices, to continue by our lifestyles to condemn these other sensibilities to the oblivion of extinction, is to rob our own senses of their integrity, and to rob our

---

58 Sarah Pink, *Doing Sensory Ethnography* (London: Sage, 2009), p. 64.
59 Yi–Fu Tuan, *Space and Place*, p. 18.
60 Soper, "Alternative Hedonism…," p. 572.
61 Pink, "Sensing Cittáslow…," p. 66.

own minds of their coherence. We are human only in contact, and conviviality, with what is not human.[62]

Seen from the perspective of phenomenology, perception is always reciprocal and dialogic. As such, read in the light of ecophenomenology, perception of the environment reestablishes the broken connection with the natural world by engaging the perceiving self in "silent conversation"[63] with the perceived, thus allowing for the reconstruction of the integrity of the senses and illustrating the ways in which "self-identities and environments are coproduced."[64] The vision of the world sought by the advocates of slow tourism as exemplified by *Slow London* rests upon Abram's "other sensibilities" which can be seen as helping to reconstruct the connection long lost in the process of alienation of our bodies from the natural world. Following Abram's rereading of Maurice Merleau-Ponty's phenomenology of perception it is possible to adopt Abram's idea of perception as a "dynamic relationship in which the perceiving body is constituted in the act of perception."[65] Such an approach to sensuous immersion in the world corresponds with the stress put on the relation between sensuous perception and the acts of self-identification/self-creation by urban designers and social anthropologists.

In response to popular representations of environmentalists as people who force themselves to live in a state of constant self-denial, Richard Kerridge asks a fundamental question: "[c]ould we come to act not only because we have been persuaded of the need, but because our physical impulses and appetites lead us that way?" and he provides an immediate answer: "[a]cting to avert climatic change would then not be a matter of restraining our impulses but of releasing them."[66] Seen from this angle, the turn towards nature is represented as stemming from a desire to reconnect with the natural world, which is possible through "a reorientation of the senses."[67]

The English poet Jeremy Hooker has provided environmentalists and ecocritics with a very useful and hence often-used metaphor of "ditch vision,"[68] whose

---

62   David Abram, *The Spell of the Sensuous* (New York: Vintage, 1997), p. 22.
63   Abram, *The Spell of the Sensuous*, p. 52.
64   Pink, "Sensory City…," p. 61.
65   Kerridge, "Green Pleasures," p. 135.
66   Kerridge, "Green Pleasures," p. 134.
67   Kerridge, "Green Pleasures," pp. 134–135.
68   See: Werner Bigell, "Distinction but not Separation: Edward Abbey's Conceptualization of Nature" (PhD. diss, University of Tromsø, 2006), p. 52, accessed 8 February 2014, http://munin.uit.no/bitstream/handle/10037/1227/thesis.pdf?sequence=1

essence consists in the ability and readiness of the perceptive mind to see the unusual in the familiar, the mode of perceiving reality which came to dominate his childhood, which he spent in the rural England. Following this mode of perception it is clear that

> natural spaces do not necessarily have to be large wilderness areas, they could even exist inside a city, and that fields, when viewed from ditches, can appear to be endless. What is important about these spaces is not their extension but the perceived difference of social control and the possibility for alternative perspectives, in other words their openness for signification and individual appropriation.[69]

Slow London takes up this challenge and makes possible the quest for the unknown in an environment which seems to be all-too-well known. What it reveals is that the other dimension is always there; hidden, not exposed to immediate consumption, requiring effort to be made, not imposing itself ostentatiously, yet already there as long as the fact of "our sensuous immersion in the world"[70] is acknowledged. An excerpt from *Slow London* may serve as an illustration of the ways in which the urban environment is transformed in the act of "individual appropriation" into an alternative space. This is how one of the authors of the book describes her encounter with the natural world in Morden Hall Park:

> An irresistible giant oak launched me into the tour: its trunk had branched into two, one half showing off familiar rough bark, the other cloaked shyly behind dried, wooded vines. Like light and dark, the tree had formed a perfect dichotomy of texture. This set me off, barefoot on the dewy grass under the full force of the sun. [...]
>
> A fallen tree became an, admittedly slightly lumpy, bench but it did the job as I watched a duck dart and dive in the river and wondered what it must feel like to skim water. My attention drifted to a tree that had been allowed to die naturally, a rare thing these days [...] (*SL*, 130).

A walk through one of London's numerous parks becomes a form of exploration that "probe[s] the jungles" (*AMFST*, 13) of the local environment. In a manner that can be associated with Annie Dillard's *Pilgrim at Tinker Creek*, an account of Thoreau-like life lived for a year in the Blue Ridge mountains in Virginia, the author opens herself up to the natural environment and lets it display to her its "textures, sounds and shapes." The description of the park resonates with Dillard's belief that "[t]he secret of seeing is, then, the pearl of great price. [...] But although the pearl may be found, it may not be sought. The literature of

---

69 Bigell, "Distinction but not Separation," p. 52.
70 Kerridge, "Green Pleasures," p. 136.

illumination reveals this above all: although it comes to those who wait for it, it is always, even to the most practiced and adept, a gift and a total surprise."⁷¹ Dillard's emotional relationship with the natural environment, framed by the awareness of her being an element of something bigger, is also similar to what David Lindo, aka The Urban Birder, states: "[y]ou lose part of your humanity if you get disconnected from nature. It's vital for sanity. I'm always birding, it's part of my spirit, it's grounded me. I come here every morning and I feel so privileged to be part of a greater process" (*SL*, 45). Curious as the connection may seem, *Slow London* can be read as a sequel to Dillard's ambitious enterprise, produced by an alternative lifestyle and embraced by the rhythm of contemporary urban everyday life; it may be viewed as Dillard's "project" in miniature, performed in an urban setting.

For David Lindo, Hooker's metaphor of "ditch vision" might be a useful tool in the process of appropriation of the natural environment and turning it into a realm of discovery and adventure. He says: "[y]ou have to imagine that this is a wilderness, that people don't exist and the buildings are the cliffs. There's more to life in cities than people realise" (*SL*, 42). The approach requires reorientation to the natural and although performed in the urban setting it still can be achieved even if by "more purely imaginative means. 'Ditch vision' names the imaginative habit of playing with scale in order to discover wildness and infinity in small spaces; the genre of daydreaming that sees in an overgrown railway bank the principle and possibility of wildness."⁷² *Slow London*, "the genre of daydreaming," promotes "the imaginative habit of playing with scale," which in the act of defamiliarization transforms the urban space, apparently well known and constitutive of the landscape of the everyday, into the space of the unknown, a space that needs to be rediscovered.

The phenomenon of slow tourism inscribes itself in the context marked by John Urry's analysis of the tourist's gaze in the sense that a slow traveller can be described as a semiotician who performs the act of "reading the landscape for signifiers of certain pre-established notions [...],"⁷³ or, pushing the analysis further, as an eco-semiotician who engages the senses in the act of reading the natural environment and lets them create a wide spectrum of "senscapes." However, it seems that slow travel also destabilizes the opposition between the

---

71 Annie Dillard, *Pilgrim at Tinker Creek* (New York: Harper's Magazine Press, 1974), p. 33.
72 Kerridge, "Green Pleasures," p. 133.
73 John Urry, *The Tourist Gaze. Leisure and Travel in Contemporary Societies* (London: Sage Publications, 1990), p. 12.

ordinary and the extraordinary, which Urry sees as a fundamental condition of tourist experience:

> Tourism results from a basic binary opposition between the ordinary/everyday and the extraordinary. […] [P]otential objects of the tourist gaze must be different in some way or other. They must be *out of the ordinary* [emphasis original]. People must experience particularly distinct pleasures which involve different senses or are on a different scale from those typically encountered in everyday life. There are many different ways in which such a division between the ordinary and the extraordinary can be established and sustained.[74]

In the case of the slow tourism experience, as illustrated by *Slow London*, the presence of the extraordinary is not the effect of the introduction of the division into work time and leisure time, which thus allows for the experience of the "out-of-ordinary," but is represented as a natural extension or continuation of everyday and ordinary practices. Since slow travel is described by Gardner as "a state of mind," the suggestion is that what underlies a slow traveller's sensitivity is the erasure of the opposition between the ordinary and the extraordinary. A slow traveller is represented as the one who seeks the extraordinary not only "outside the normal places of residence and work"[75] but in the everyday, the mundane and the familiar.

Conclusions

Although it seems tempting to look for the pioneers of slow travel in ancient Rome or among medieval pilgrims or participants in the Grand Tours, the question whether slow travel has its antecedents in past forms of travel remains disputable. To see today's slow travel in terms of historical continuity is to reduce it to a one-dimensional phenomenon whose essence consists in a preference for slow modes of transport over fast ones. Slow travel is a phenomenon whose ideological ramifications cannot be limited to a preference for slower modes of transport, but, as a dialectical phenomenon, it should be viewed in the context of its opposite, lest its contesting and resistant character be eclipsed. Slow travel is about agency and individuality in an age of mass production of cultural practices and meanings and it is only in contrast with the mainstream models that the ideological framework of slow travel can display its subversive potential. As Urry claims:

74　Urry, *The Tourist Gaze*, pp. 11–12.
75　Urry, *The Tourist Gaze*, p. 3.

Slow travel [...] represents a way of consumer thinking about tourism, where there is synergy between experiential aspects of travel and, for some tourists, discourses about the environment, in particular climate change. By travelling slowly, people are not just choosing a mode of transport, but they are also negotiating with place, the environment, their personal identity as a tourist, and in some cases, expressing certain ethical and ideological values.[76]

Urry's study of leisure and tourism leaves no doubt as to what forms the foundation of tourist experience: the gaze. As he says: "[...] we gaze at what we encounter. And this gaze is as socially organised and systematised as is the gaze of the medic."[77] Urry's approach is shaped by the conviction that the tourist's gaze is a construct which has its own history which reflects evolving patterns of cultural practice. Following this mode of thinking about the constructed nature of the tourist gaze, it is possible to reproduce slow travel as a new object of the tourist gaze, the fact that illustrates "changing [...] distinctions of taste within the potential population of visitors."[78]

What most slow travel publications have in common is that their rhetoric is marked by a logic of binary oppositions, according to which what is slow is deeper, easier and more rewarding than what is not slow, the logic being emblematic of what might be seen as *slowwashing*. Slowness is seen as a remedy and recipe for change: slow travel emerges as a means of metamorphosis which is very likely to happen when a tourist spends a week living with the locals and following the rhythm of life of a place he has chosen to spend his slow time in. Having spent a week or more in this way, the slow traveler is expected to "come away from [...] (a) trip rejuvenated and changed"[79] though the quality of change is never specified. The change is taken for granted as part and parcel of the influence of slowing down which is illustrative of the huge transformation the very concept of slowness has undergone in the last two decades. Slowness becomes associated with change and, which is often implied, with spiritual rebirth.

Bill Newlin, the Avalon Travel publisher, observes that "[t]hey [contemporary tourists] are valuing time over money, looking for ways to make educated decisions. People want to find something new, have stories to tell, but what that means has changed. [...] The unknown is harder to find today but the craving for

---

76 Dickinson, Lumsdon, Robbins, "Slow Travel…," p. 282.
77 Urry, *The Tourist Gaze*, p. 1.
78 Urry, *The Tourist Gaze*, p. 4.
79 "What is Slow Travel?"

adventure survives."[80] Since the world has shrunk radically and there are hardly any unexplored places left, the sense of adventure undergoes redefinition and entails the rise of a novel approach to places which are already well known on the surface but are still capable of offering the possibility of diving deep beneath what is customarily highlighted by traditional tourist description, satisfying travellers' hunger for novelty. The object of tourist exploration does not have to change; what changes is the mode of tourist experience.

---

80   Bill Newlin, quoted by Carl Parkes, "The History of Travel Guide Books," accessed 27 January 2014, http://travelwriters.blogspot.com/2006/04/history-of-travel-guidebooks.html

Chapter Seven: "Relocation, Relocation": Slow Goes Pop

> There is clarity in this Tuscan life.
> Marlena de Blasi,
> *A Thousand Days in Tuscany*

> Here, I seem to go from pocket to pocket of calm,
> quiet, broken only by the beautiful warble of a wren,
> or the singing of skylarks.
> Tessa Hainsworth,
> *Up with the Larks. Starting Again in Cornwall*

Chris and Gillean Sangster, the authors of *The Downshifter's Guide to Relocation: Escape to a Simpler, Less Stressful Life*, one of numerous popular guides offering advice to those who long for a seachange in their lives, write in the "Preface":

> There is an ever-increasing interest in the prospects of both downshifting and relocation, demonstrated through the number of newspaper and magazine articles, television programmes and familiarization workshops. The pressures of work and the stresses and strains of city life have raised the question in many people's minds as to whether there is an alternative way of living.[1]

Recently the question "whether there is an alternative way of living" has been positively answered by slow living. Having evaluated slow living in the context of its marketability, popular culture has adopted some of the ideas of slow, sustainable living and turned them into products which address a popular audience, i.e. not specialized groups of people with slow-oriented tastes but audiences "large and diverse" which "would not necessarily have a green or anti-capitalist orientation"[2] and whose main objective is not so much to popularize selected aspects of slow living philosophy but to sell well. Popular literature, along with lifestyle television, turns out to be one of the most useful popular products in this respect, a vehicle for popular conceptions (and misconceptions) about what it means to make an attempt to reinvent one's relationship with time in a rural environment.

---

1 Chris and Gillean Sangster, *The Downshifter's Guide to Relocation: Escape to a Simpler, Less Stressful Life* (Oxford: How To Books Ltd, 2004), p. xi.
2 Lynn Thomas, "Alternative Realities. Downshifting Narratives in Contemporary Lifestyle Television," *Cultural Studies*, Vol. 22, No. 5 (2008), p. 681.

This chapter examines narratives of slow life in popular literature from the perspective of the ways in which they represent a dissatisfaction with lifestyles promoted by the culture of speed and flow, and a new, more balanced life led outside the urban rat-race. They are approached as popular articulations of concerns with a sense of disappointment with the ways contemporary life develops and dreams about alternative reality. Often framed by black-and-white rhetoric, they mirror, in a manner not free from simplification and idealization, the anxieties and dilemmas of a mis-man whose identity has been marked by accelerated speed of life in the culture of excess.

The novels analyzed in this chapter have been divided into two categories. Read from the perspective of revalorization of slowness, they exemplify the ways in which the discourse of slowness has inseminated two literary trends: the trend of an idealization of Italy in Western literature and a literary genre, chick lit. The first category of relocation narratives is associated with the so-called "at home in Tuscany"[3] novels, the term used by Wendy Parkins in her analysis of novels set usually in Mediterranean Italy, which she interprets within the framework of slow living philosophy and relations between cosmopolitanism and domesticity. The second category, which I propose to call *slow lit*, includes humorous accounts of country living written in the manner characteristic of chick lit. The novels in both categories provide audiences with insight into the everyday experience of slow life outside the urban rush which is often nostalgically and idealistically (Tuscan novels) or humorously (slow lit) romanticized. They also construct a mythology of deceleration which provides narratives that point toward the possibility of alternatively defined pleasure, meaning and happiness. The protagonists are no longer "temporary identities on vacation,"[4] to refer to Karen Stein's study, but characters who search for the good life in the country and for whom slowing down is much more than just a passing fad: it is a lifestyle which brings about mindfulness and attentiveness and which helps reconstruct time horizons so that neither the past nor the present are "threatened by the tyranny of the moment."[5]

---

3 Wendy Parkins, "At Home in Tuscany: Slow Living and the Cosmopolitan Subject," *Home Cultures*, Vol. I, Issue 3, (2004), p. 258.

4 Karen Stein, "Getting Away from It All: The Construction and Management of Temporary Identities on Vacation," *Symbolic Interaction*, Vol. 34, Issue 2 (2011).

5 Thomas Hyllard Eriksen, *Tyranny of the Moment. Fast and Slow Time in the Information Age* (London: Pluto Press, 2001), p. 2.

1. "[T]o the measure of a (slow) man": Alternative Hedonism under the Tuscan Sun

The landscape of popular culture is increasingly informed by a quest for "alternative realities."[6] The recent explosion of interest in slow living has brought about a wide range of cultural products and practices which embrace the idea of deceleration as a path to a reality whose foundations are constructed through a redefinition of time and speed. Lyn Thomas' analysis of British lifestyle television programmes is illustrative of the present popularity of the ideas of slowing down and "desires to change lives and their realization."[7] The programmes analyzed by Thomas include, among others, *River Cottage* (1999–) and *Relocation, Relocation* (2003–2011), two British television programmes inspired by a popular conflation of the good life with life in the country. As Thomas explains, they are both based "on the search for the dream life, location and house, and [...] are all premised on the need for escape from what are perceived as negative aspects of contemporary, and overwhelmingly urban, life."[8]

The motif of relocation has been recognized for its appeal to mass audiences not only by television but also by the European book market. One can observe the proliferation of narratives in which the plot is constructed along the lines of a contrast between the country and the city and whose protagonists change their lives, so far led in a large, busy city, by settling in the country. What these have in common is that in most of the cases they are deeply rooted in a desire to leave behind the neurotic rhythms of globalized postmodern culture which the city exemplifies. The purchase of an old house in the south of Europe – Italy, France, or Spain – becomes an element of the rite of passage from the world of excessive consumption, accelerated lifestyle and anonymity of the urban crowd to the world of sustainability, slowness and conviviality. Popular examples of such narratives, often a combination of memoirs, travelogues and food literature, are, to name but a few, *Under the Tuscan Sun* (1996), *Bella Tuscany* (1990) and *In Tuscany* (2000) by Frances Mayes, *The Hills of Tuscany* (1998) by Ferenc Mate, *A Year in Provence* (1989) by Peter Mayle and *Driving Over Lemons: An Optimist in Andalusia* (2009) by Chris Stewart. Although successful in commercial terms, this type of novel remains largely ignored by literary and cultural analysts. An interesting study has been carried out by Wendy Parkins, whose analysis of the

---

6   Thomas, "Alternative Realities…," pp. 680–699.
7   Thomas, "Alternative Realities…," p. 685.
8   Thomas, "Alternative Realities…," p. 686.

representation of "a particular kind of 'slow' subject"⁹ in Tuscan novels remains the main point of reference for anyone wishing to explore the issue of literary representations of slow living in writing today. The analysis presented here aims to examine the representation of alternative hedonism in *A Thousand Days in Tuscany* (2005) by Marlena de Blasi, a motif which seems to warrant critical attention due to its correspondence with the counter-cultural and subversive potential of deceleration.

Parkins' analysis of Tuscan novels challenges the image of homogeneous Tuscan fiction capable only of almost automatic repetition of fixed narrative conventions. One of the first novels to set the conventions for the "Tuscan farmhouse literature"¹⁰ and popularize the idea of relocation as an answer to a growing sense of disappointment with the accelerated character of the present is Peter Mayle's *A Year in Provence* (1989). However, as Parkins is quick to notice, "Mayle's account is charmingly free from the realities of everyday life, like earning a living: the only drudgery he and his wife seemed to experience was cleaning up after household guests and sweeping up after tradesmen."¹¹ A similar representation of Tuscany may be found in the novels of Frances Mayes, who builds the images of a "carnival community" which Parkins contrasts with an "ethical community" in the novels of Ferenc Mate.¹² She goes on to claim that while the protagonists of Mayes' novels are visitors who settle in Tuscany with the intention of returning to their homeland at some point of their lives, and thus constantly fetishize the community they visit, the characters in the novels by Mate undertake an effort to become legitimate members of the communities they live in rather than visit, and to understand the ethical principles which organize those communities' lives.

As has been noted earlier, relocation novels are often set in Italy, which enjoys a special status in the European geography of slowness. The fact that this is where the Slow Food and Slow Cities movements started is to some extent determined by the fact that in Italy "slowness has been a more readily available cultural resource."¹³ Italy has always been an object of desire for anyone ready to leave behind the modern rat-race and a task-oriented lifestyle; its long "history of functioning as an idealized site outside modernity (for non-Italians) where the

---

9   Parkins, "At Home in Tuscany," p. 258.
10 Parkins, "At Home in Tuscany," p. 257.
11 Parkins, "At Home in Tuscany," p. 259.
12 Parkins, "At Home in Tuscany," p. 264.
13 Wendy Parkins, Geoffrey Craig, *Slow Living* (Oxford: Berg, 2006), p. 83.

self can be rediscovered and refashioned"[14] has been enriched by recent popularization of the country by popular culture. However, what needs to be emphasized is that similar emotions have been invested in other geographical parts of Europe, thus allowing one to draw a map of literary and cultural deceleration where a central position is occupied by Tuscany but where Provence in France, Andalusia in Spain and Cornwall and Scotland in Great Britain are also marked for their potential to offer conditions for a "life of pleasure, tranquillity, soul-restoring physical labour and harmony with nature,"[15] and to become popular relocation sites.

"Look at that Tuscan landscape. This is where everyone in the world would like to live,"[16] says Marlena de Blasi, the author of *A Thousands Days in Tuscany*, to herself, and her words are emblematic of the ways in which popular culture has recently claimed Tuscany as a site of good, sustainable living and commodified its image. Ezra Pound in *The Spirit of Romance* explored the specificity of Provence and Tuscany and claimed that the cult of Provence was a cult of emotions, which led to the rise of a psychology of emotions, while what dominated in Tuscany was "a cult of the harmonies of the mind."[17] As Parkins observed, a sense of balance and harmony permeates the images of rural idyll in Tuscan novels:

> Tuscany figures as a utopian space in which, for the price of a farmhouse and land (and usually extensive renovation), it is possible to live the good life [...]. Fundamental to this utopian space called Tuscany [...] is an evocation of slow living – *la vita lenta* – based on proximity, honesty, simplicity, and, perhaps most importantly, conviviality, which imaginatively resolves the contradictions and dislocations of postmodern culture.[18]

Relocation novels are products of a sense of anxiety experienced by those who are tired with the pace of life and its consequences. A change of place and home is either the result of changes in the lives of main protagonists provoked by various unwanted circumstances (e.g. divorce), or the result of a desire to liberate oneself from the pressures of urban life. "I smile at the aptness of the bridge's name. Liberty. What better road for an escape?" (*ATDiT*, 1) de Blasi asks herself while driving with her husband to Tuscany and heading for a life which she describes to herself as a "rebellion against structure" (*ATDiT*, 62).

---

14 Parkins, "At Home in Tuscany," p. 256.
15 Parkins, "At Home in Tuscany," p. 258.
16 Marlena de Blasi, *A Thousand Days in Tuscany* (London: Virago, 2008), p. 23. Throughout the chapter the book will be referred to as *ATDiT*.
17 Ezra Pound, *The Spirit of Romance* (New York: New Directions, 2005), p. 116.
18 Parkins, "At Home in Tuscany," p. 258.

Regardless of the motives behind the moves, relocation narratives are buttressed by a belief in the power of a new beginning. "A desire to live differently through both a change of location and a change in temporality"[19] resonates with Kate Soper's alternative hedonism in many ways. A redefined concept of the good life is rooted both in a sense of disappointment with the promises made by consumerism and a conviction that it is possible to promote "new modes of thinking about human pleasure and self-realization"[20] which may be seen as strategies for resistance against the homogenizing practices of consumer culture.

Such a mode of thinking, which demands, as Soper puts it, being "more assertively utopian in promoting sustainable consumption,"[21] allows one to establish a connection with Olga Tokarczuk and her proposal for the social game she calls heterotopia, the idea of which consists in an attempt to imagine the world as a place operating along radically different lines. For Tokarczuk social organization of our lives is the product of a certain intelectual inertia.[22] Once this has been overcome, once the assumptions of social, political and cultural life have been questioned, suspended and redefined, once we have crossed the current borders of our imagination, we will be able to enter the alternative reality of heterotopia – a world constructed around a set of differently defined principles. What we take for granted as logical and natural, in the imaginary land of Heterotopia is approached with scepticism, sometimes incredulity, often laughter. Such are the reactions of the Heterotopians to the ideas of gender division into femininity and masculinity (they think about gender continuum rather than gender polarization), family defined by blood relations (for them family membership is based on individual choice), human mobility seen as a process to be monitored (to be able to move freely is one of the most fundamental privileges of every Heterotopian), childbearing seen as a human right (in the face of overpopulation and world hunger, giving birth is something that has to be controlled in Heterotopia) or religion understood as the source of legislation (in Heterotopia religion is as private as one's sexual life).

---

19 Parkins, "At Home in Tuscany," p. 257.
20 Kate Soper, "Alternative Hedonism, Cultural Theory and the Role of Aesthetic Revisioning," *Cultural Studies*, Vol. 22, No. 5 (2008), p. 571.
21 Kate Soper, "The Other Pleasures of Post-Consumerism," *Soundings. A Journal of Politics and Culture*, Issue 35 (2007), p. 35. Accessed 15 August 2013, http://www.lwbooks.co.uk/journals/soundings/issue/35.html
22 Olga Tokarczuk, *Moment niedźwiedzia* (Warszawa: Wydawnictwo Krytyki Politycznej, 2012), p. 14.

Proposed as an exercise in intellectual vigour and readiness to undermine the premises of social organization, rather than a call to build a utopian society, Tokarczuk's heterotopia allows us to read Soper's concept of alternative hedonism, which can "form desire and encourage a different structure of feeling and affective response to the world of material culture,"[23] as an example of a "heterotopian antithesis."[24] The purpose behind highlighting the correspondence between Soper's socio-cultural reading of consumer culture and Tokarczuk's essay is not to disparage the power of Soper's concept by relocating it to the realm of pure utopia but rather to illustrate the possibility of a sustainable life in conditions of unsustainable over-consumption, provided that the assumptions of dominant lifestyles cease to be seen as irreversible. What both Soper and Tokarczuk suggest is the possibility of an alternative reality, which does not have to be dismissed as purely utopian but can be created and negotiated in everyday practices.

A dissatisfaction with the present that awakes a desire to refashion one's self may take various forms:

> It may be a nostalgia for certain kinds of material, or objects or practices or forms of human interaction that no longer figure in everyday life as they once did; it may be the case of missing the experience of certain kinds of landscape, or spaces (to play or talk or loiter or meditate of commune with nature); it may be a sense that possibilities of erotic contact or conviviality have been closed down that might otherwise have opened up; or a sense that were it not for the dominance of the car, there would be an altogether different system of provision for other modes of transport, and both rural and city areas would look and feel and smell and sound entirely different. Or it may be just a vague and rather general malaise that descends in the shopping mall or supermarket: a sense of *a world too cluttered* [emphasis mine] and encumbered by material objects and sunk in waste, of priorities skewed through the focus on ever more extensive provision and acquisition of things.[25]

Suspended between the sense of excess, a product of insatiable consumer greed constantly fuelled by ideologies provided by dominant models of consumer lifestyles, and the sense of loss, a post-consumer is now more likely than ever to start to negotiate the conditions of living in what not so long ago seemed to be a consumerist paradise. De Blasi and her husband abandon Venice prompted by a growing disaffection with the unbalanced and disharmonious life they lived there. Her declaration that the reason behind coming to live in a small Tuscan

---

23 Soper, "The Other Pleasures of Post-Consumerism," p. 35.
24 Tokarczuk, *Moment niedźwiedzia*, p. 5. Translation mine.
25 Soper, "The Other Pleasures of Post-Consumerism," p. 35.

village is to "make a life scrubbed clean of clutter, a life that follows the rhythms and rituals of this rural culture" (*ATDiT*, 36) is at the same time a criticism of urban culture and glorification of rural culture as a repository of healing values and traditions. De Blasi shares with Soper and other advocates of alternative hedonism a sense of living in "a world too cluttered;" chaos, confusion and disarrangement are contrasted with clarity and orderliness. The Tuscan village de Blasi and her husband have chosen for their new life becomes a symbol of their quest for meaning in the world of flux and renting a villa is a part of a rite of passage from the world of clutter to the world of clarity.

Their criticism of an accelerated life in the city is deeply embedded in their disappointment with the promises of progress, the side-effects of which have come to dominate their lives. "Our suspicion is strong that there is a greater peace in going backwards" (*ATDiT*, 36), she says and it is clear that she sees the character of the rural-urban divide as formed by the logic of binary oppositions. In the light of this logic the type of life led by Tuscan villagers is "made *a misura d'uomo*, to the measure of a man" (*ATDiT*, 36), a concept which resonates with the ideas of sustainable living, the rhetoric of which has become increasingly popular in recent cultural discourses. Seen from this perspective, urban life turns out to be the opposite of proportion, balance and harmony.

The line of binary opposition is constantly reinforced; it often takes the form of explicit lectures on the nature of difference between the two lifestyles. Barlozzo, the village's wise man, who becomes a teacher of the newly arrived couple, illustrates the difference by referring to two groups of people in the village, two "social sects" which include "*i progressisti* […] chomping to leap into the future" and "*i tradizionalisti* [who] court the rituals, saying that the only true progress waits a few steps back into the past" (*ATDiT*, 77).

What helps to accentuate the difference is the attitude to objects, artefacts of everyday life. *I progressisti* are those who crave the comfort and effortlessness of the apartments in the lower town where there is "[n]o more carrying wood up the stairs and ashes back down the stairs, who are longing for built-in, plastic-finished closets rather than those cherry-wood armoires, big and deep as caves and who want great, fake suns swaying from the ceilings instead of the rough, hand-wrought iron lanterns Biagiotti's grandfather forged for the whole village a hundred years ago" (*ATDiT*, 77). A dream about a condominium represents a departure from tradition, which is perceived as the source of meaning and continuity. For *i progressisti* the value of an object consists in its utility and is not determined by its historical aspect and the role it plays in the process of the creation of a sense of cultural continuity. Objects, when seen as parts of personal narratives, reveal their potential to help form the identity of a home and its inhabitants. Once the relationship with an object is established, what takes place is the "transposition of temporality" that Arjun Appadurai explains as "the subtle shift of patina from the object to its owner or neighbour" which is then followed by the fact that "the person (or family, or social group) himself or herself takes on the invisible patina of reproduction well managed, of temporal continuity undisturbed."[26] "Fake suns" become symbols of the replacement of uniqueness by mass production and separation from historical continuity, which is manifested in disregard of the cultural values of traditionally produced objects, while hand-wrought iron lanterns tell a story of attachment, rootedness and cultural perseverance.

What *i progressisti* disregard, *i tradizionalisti* embrace. Here is what Barlozzo says about them:

> They say life was better when it was harder. They say food tasted better laid down over hunger and that there's nothing more wonderful than watching every sunrise and every sunset. They say that working to the sweat, eating your share, sleeping a child's sleep is what life was meant to be. They say they don't understand this avid bent to accumulate things you can't eat or drink or wear or use to keep you warm. They remember when accumulation still meant gathering three sacks of chestnuts instead of two. They say their neighbors have lost the capacity to imagine and to feel – some of them, the capacity to love. They say since we all have everything and we all have nothing, our only task is to keep searching to understand the rhythm of things [...]. Live gracefully in plenty and live gracefully in need. [...] Ease and plenty seep together to form a single sentiment that comes out looking a good deal like nonchalance. All that ease, all that plenty. What can one expect of them but nonchalance? (*ATDiT*, 79–80).

Barlozzo is represented as an objective observer who takes no sides and whose sympathies do not affect his matter-of-fact narrative. Both *i tradizionalisti* and *i progressisti* are "they" to him, a pronoun which establishes a sense of distance and is supposed to be a guarantee of objectivity regardless of the fact that the life he leads unmistakably places him among *i tradizionalisti*.

One may notice a certain correspondence between Barlozzo's exposé and John Tomlinson's principle of immediacy which he sees as a motive force of contemporary culture. For Tomlinson a turn towards effortlessness is characteristic of the cultural practice of post-industrial societies. What Barlozzo describes as "*sprezzatura*. [...] 'the state of effortlessnes,' [...] the mastering of something – an art, a life – without really working at it, with the result being nonchalance" (*ATDiT*, 80) resonates with Tomlinson's analysis of the ways in which mechanical speed has been reinventing social and individual practice. "[A] displacement

---

26 Arjun Appadurai, quoted by Parkins in "At Home in Tuscany," p. 262.

of effort and labour from the terms of the popularly imagined social contract"[27] has generated conditions in which a definition of the good life assumes the absence of effort which has been turned into a marker of discomfort and failure. For Tomlinson, the culture of speed is the culture of "legerdemain, the sleight of hand," which, he claims, "can be understood as a lightness of touch and as a mystification."[28] While Tomlinson inscribes lightness of touch in the context of the rise of new technologies of communication, which require the use of the hand from anyone operating keyboards or using touch pads or handsets, the aspect of mystification introduced by the concept of legerdemain is related to

> the apparent ease with which, in developed economies, and for larger sectors of the population within these economies, goods, and particularly technological advances, seem simply and continuously to be delivered, become affordable, morph from luxuries into necessities.[29]

The affordability of mass-produced goods, these "fake suns swaying from the ceilings," as well as the ubiquity of services and the speed of their delivery, create the foundations of the culture of immediacy, which, as we remember, Tomlinson described as marked by the disappearance of the gap between a desire and its fulfilment. The difference between Barlozzo's *i progressisti* and *i tradizionalisti* can thus be found in the latters' valorization of effort and readiness to redefine the concept of "the achievable good life."[30]

What reinforces a sense of difference between the city and the village is that the latter seems to be unaffected by the present. Taking part in a spontaneously organized welcome party and observing its participants, de Blasi asks herself: "1920? 1820? How is this evening different from an evening in June when the oldest man here was young, I wonder" (*ATDiT*, 19). "[L]ost in time" (*ATDiT*, 23), the village takes its vitality and energy from a sense of continuity and the re-enactment of past practices and rituals which, when repeated, demonstrate their power to unite and consolidate. The memory of the place is refreshed daily and the practices of everyday life function as a vehicle of communal memories and the memory of the village's rich past. Baking grape flatbreads is one of many occasions to emphasize the cultural perseverance of the place: "[s]ee. Surely the Etruscans also made this bread during the harvest.

---

27   John Tomlinson, *The Culture of Speed. The Coming of Immediacy* (Los Angeles: Sage, 2007), p. 80.
28   Tomlinson, *The Culture of Speed*, p. 80.
29   Tomlinson, *The Culture of Speed*, p. 81.
30   Tomlinson, *The Culture of Speed*, p. 80.

It's an ancient thing." (*ATDiT*, 107). The sense of "ancestral energy" (*ATDiT*, 210) permeates everyday life in the village and binds people together. It is represented as a real community where the communion of its members is not imagined but lived on a daily basis.

An individual in de Blasi's story always remains in a relation to the community he is a part of. In return for their neighbours' offer to clean the house de Blasi and her husband rent, the couple invite everyone to a local bar "for *aperitivi*" (*ATDiT*, 14), a social event which naturally and spontaneously grows to become a sort of a village festival: "[m]ore tables are unearthed from the cellars of the nearby city hall and soon the whole piazza is transformed into an alfresco dining room" (*ATDiT*, 15). The naturalness with which the villagers organize themselves to dine together is emblematic of the strength of social bonds that bind them together. It is illustrated by their willingness both to embrace the pleasure of being part of the community through eating locally and organically produced food, and to stress their sense of the ownership of public space: "they move easily between their private spaces and the public domain of the piazza. *Both belong to them*" [emphasis mine] (*ATDiT*, 15). While globalization increasingly promotes processes and practices which stem from "disengagement from specific, local contexts,"[31] the villagers' relations with space are defined by the experience of the locality as the site of meaning and continuity.

A sense of alternatively defined pleasure is often informed by a redefinition of one's relation with public spaces, which offers their users a chance not only to escape from "a constricted pattern of daily life"[32] but also to liberate themselves from overdependence on the places of home and work. We remember that Ray Oldenburg's third place, a site of interactions between people which are different from the ones carried out in the first two places, is a neutral terrain where the relations of power typical of home and work do not operate, and which is defined by a sense of comfort, equality and conviviality. Watching the villagers who are enjoying themselves in this third place, de Blasi observes:

> Many here seem beyond fifty, some twenty or thirty years more. Those who are younger echo their elders' kindness and somehow seem older than their years. There is less distinction among the generations. A girl of perhaps seventeen gets up to fix a plate for her grandmother, telling her to watch for the bones in the rabbit stew, asks her if she's taken her pills. A boy no more than ten, slices the bread, telling his younger brother to stay clear of his work, that he should never play where someone is using a knife (*ATDiT*, 18–19).

---

31   Parkins, Craig, *Slow Living*, p. 70.
32   Ray Oldenburg, *The Great Good Place* (Cambridge, Mass.: Da Capo Press, 1999), p. 9.

Performing at the table their everyday routines, the villagers offer an insight into the ways in which "networks of meaning and pleasure"[33] construct culture. De Blasi sketches an idealized image of a rural community whose stability derives from the sense of co-responsibility and reciprocity. What characterizes urban communities is that in most cases cross-generational relations and bonds have been weakened by the mobility of people, which, as Tim Creswell puts it, "appears to involve a number of absences,"[34] and their increasing detachment from what Edward Relph calls "significant places."[35] In the village de Blasi writes about, which is an example of a significant place for its inhabitants, the relations between family members and neighbours are maintained naturally and on a daily basis and therefore the place functions as a site of meaning, which is informed by the sense of broadly understood attachment, involvement and commitment.

Among the variety of practices performed daily by the inhabitants of the village which help turn it into "a center of meaning and field of care"[36] those which revolve around food play a special role in the production of meaning. Parkins and Craig are careful to note that "Tuscany is reconfirmed as a utopian space of rural simplicity and convivial pleasures in the Anglo imaginary and the discovery of cultural authenticity in inextricably linked with food."[37] In the Tuscan village food production and the celebration of food-oriented pleasures are at the heart of social and communal life. Barlozzo explains that next to the weather, which affects every aspect of a farmer's life, and death and birth, food is the third subject worth discussing:

> [...] all three of these major subjects encompass, in one way or another, philosophy, psychology, sociology, anthropology, the physical sciences, history, art, literature and religion. We get around to sparring about all that counts in a life but we usually do it while we're talking about food, it being a subject inseparable from every other subject. It's the table and bed that count in life. And everything else we do, we do so we can get back to the table, back to bed (*ATDiT*, 32).

The figure of Barlozzo can be read as an embodiment of traditional wisdom and knowledge the loss of which has been caused by the unimaginative development of economies and the domination of accelerated lifestyles. What he says seems to

---

33 Graeme Turner, *British Cultural Studies. An Introduction* (London: Routledge, 2003), p. 2.
34 Tim Creswell, *On the Move. Mobility in the Modern Western World* (New York: Routledge, 2006), p. 31.
35 Edward Relph, *Place and Placelessness* (London: Pion, 1976), p. 1.
36 Creswell, *On the Move*, p. 31.
37 Parkins, Craig, *Slow Living*, p. 101.

resonate with the model of food-oriented culture which inspired Carlo Petrini to create the ethical, economic and cultural foundations of his Slow Food movement whose roots are to be found in the conviction that "alimentation is an essential part of life and that quality of life is therefore inevitably linked to the pleasure of eating in healthy, flavorful, and varied ways."[38]

The type of life Barlozzo introduces the newly arrived city-dwellers to places great significance upon "learning to take pleasure in the diversity of recipes and flavors, recognising the variety of places where food is produced and the people who produce it, and respecting the rhythm of the seasons and of human gatherings."[39] The villagers described by de Blasi live in a world which challenges the dictatorship of "alimentary monoculture," which Petrini describes as "the restricted range of foods and flavors experienced by those who simply accept what is most easily available."[40] The way food eaten in the village is produced is a part of "agricultural and alimentary heritage"[41] whose character is marked by the concern it places on local production and the preservation of traditional recipes:

> Fresh from their triumph of the squash blossoms, now Bice and Monica come back laden with platters of prosciutto and *salame – cose nostre*, our things, they say, a phrase signifying that their families raise and butcher pigs, that they artisinally fashion every part of the animal's flesh and fat into one sort or other of sausage or ham (*ATDiT*, 16).

Eating home-made *panna cotta* in a local bar serves as a pretext not only to look at the ways in which food is produced but also to examine the relations between human and non-human world. As de Blasi learns, the milk the dish is made of comes from Assunta, "Piero's [...] finest cow" (*ATDiT*, 21), whose uniqueness is described with passion and love by the local milkman. De Blasi observes: "[f]rom her teats to my spoon with only Pioggia's jar and Bice's pot in between. These facts redraw my concept of 'fresh.' And so I eat blue-eyed Assunta's milk, coaxed from her by a man called Rain, and it's delectable. I lick both sides of my spoon and scrape the empty bowl [...]" (*ATDiT*, 21). The opportunity to trace and identify every stage of production of food consumed every day has been removed from the experience of modern man, and the result of this separation is the loss of the whole spectrum of natural tastes and their replacement by tastes

---

38 Carlo Petrini, *Slow Food: The Case for Taste* (New York: Columbia University Press, 2001), p. xvii.
39 Petrini, *Slow Food*, p. xvii.
40 Petrini, *Slow Food*, p. 21.
41 Petrini, *Slow Food*, p. 12.

"formed by what the food industry puts before them"[42] [i.e. consumers] and the deepening of the gap between the human and non-human world. When animals are produced for food, when they become "a human product like rayon,"[43] our ability to see us and them in the context of reciprocal dependency is severely hindered.

Tuscan life is represented as successfully avoiding what David Abram calls "the structures of our civilised existence whose essence consists in our obliviousness to nonhuman nature."[44] The life led in the Tuscan village is marked by the fact that the relation between the human and nonhuman world has never been broken. We remember that for Abram "we are human only in contact, and conviviality, with what is not human."[45] The story told by de Blasi shows respect for the nonhuman world, which manifests itself in the mindfulness with which the villagers negotiate their relations with the natural environment. When harvesting chestnuts de Blasi is instructed by Barlozzo to "take only the shiniest, fattest nuts, [...] and leave the smaller ones for the beasts" (*ATDiT*, 143). This lesson of moderation and restraint as well as the related lesson of interdependency and reciprocity challenge the anthropocentrism of culture and put the spotlights on the "hidden centrality of the earth in all human experience."[46]

De Blasi's decision to "be short on comforts" (*ATDiT*, 36) and live a life without central heating, telephone and television in a small Tuscan village may have been inspired by a desire to challenge the influence of "human-made technologies that only reflect us back to ourselves."[47] The life of alternative hedonism is informed by a redefinition of key concepts which mark the direction of the development of present-day societies. The paradox is that what makes pleasure "alternative" is in fact marked by a return to what for centuries was a natural source of pleasure. Food culture is instructive in this respect for it provides a perspective from which to observe practices of the culture of excess. Writes de Blasi:

> Here poor people eat far more magnificently than the rich in America. And so I ask myself [...] – what does it mean to be poor? I think I'm learning how to live gracefully in need as well as in abundance. Essentially the approach is the same. But the trick is to define *abundance* [emphasis original]. For us abundance has become a giara of just-pressed oil, far fewer things, a little more time (*ATDiT*, 216–217).

---

42   Petrini, *Slow Food*, p. 68.
43   Annie Dillard, *Pilgrim at Tinker Creek* (New York: Harper's Magazine Press, 1974), p. 4.
44   David Abram, *The Spell of the Sensuous* (New York: Vintage, 1997), p. 28.
45   Abram, *The Spell of the Sensuous*, p. ix.
46   Abram, *The Spell of the Sensuous*, p. xi.
47   Abram, *The Spell of the Sensuous*, p. 22.

De Blasi's rehabilitation of poverty may strike a note recalling the romanticized narrative of the rural-urban divide reproduced by the wealthy middle-class in quest for novelty, because it ignores the helplessness and despair which poverty often generates. Yet when taken from the perspective of consumerism and its demands it becomes clear that de Blasi is not idealizing the state of being poor but rather criticizing the excessive character of consumption today. It is the power of counter-consumerism, which often takes the form of simplicity, that de Blasi praises in opposition to the power of excess. Her desire to live a life "scrubbed clean of clutter" resonates meaningfully with Richard Gregg's idea of voluntary simplicity, which he explains as "singleness of purpose, sincerity and honesty within, as well as avoidance of exterior clutter, of many possessions irrelevant to the chief purpose of life."[48] De Blasi's Tuscan life is motivated by a desire to live a harmonious and purposeful life in which alternatively-defined pleasure works as an organising principle. Duane Elgin, the author of *Voluntary Simplicity*, says:

> To live more simply is to unburden ourselves – to live more lightly, cleanly, aerodynamically. It is to establish a more direct, unpretentious, and unencumbered relationship with all aspects of our life: the things that we consume, the work that we do, our relationships with others, our connection with nature and the cosmos, and more.[49]

The model of alternative hedonism which de Blasi describes in *A Thousand Days in Tuscany* challenges the premises of Western "good life" and high standards of living. It involves disentanglement from what affluent culture has valorized as an absolute must and what it has promoted as indispensable for the sense of happiness. For de Blasi pleasure and simplicity are intertwined; a life led "with a minimum of needless distraction,"[50] free of clutter and excess, creates a space for mindfulness, meaningfulness and sustainability, the qualities that advocates of alternative hedonism and slow life see as indispensable for conscious negotiation of the displeasures of consumerism.

2. "Out of Your Townie Mind": The Rise of Slow Lit

Geographical location is not the main and exclusive marker of the Tuscan novel, as some of the novels that fall into this category are set in, for example, Provence

---

48   Richard Gregg, *The Value of Voluntary Simplicity* (Auckland: The Floating Press 2009), p. 6.
49   Duane Elgin, *Voluntary Simplicity. Toward a Way of Life That is Outwardly Simple and Inwardly Rich* (New York: Quill, 1993), p. 25.
50   Elgin, *Voluntary Simplicity*, p. 24.

(*A Year in Provence* by Peter Mayle) or in Spain (*Driving Over Lemons: An Optimist in Andalusia* by Chris Stewart). The category of Tuscan novel, or "Tuscan farmhouse" literature, is used by Wendy Parkins "even though some textual examples are located in other regions of Italy [...] as this is the most common and well-known manifestation."[51] What allows one to give them a single label is not so much geographical location but rather as a certain narrative pattern around which a distinctive slow life philosophy is constructed, a type of slow lifestyle led by characters, and often an idealized representation of a local community. The Tuscan novel is a literary convention of the representation of a reality characterized not only by geography but also by a certain style and narrative tone, which are informed by an assumption that Italy (or Spain or France) can be seen as "an Arcadian backdrop against which the existential dilemmas of modernity could be illuminated and explored by subjects from elsewhere."[52]

Popular as it is, the Tuscan model of relocation narratives is not the only one that disseminates among a wide readership the idea of "changing place to change pace," to refer to an analysis of the pace of places by Jenny Shaw.[53] A longing for an alternative reality and alternative sources of pleasure in a world whose character has been increasingly marked by a sense of consumerist displeasure, has been recognized and embraced by British chick lit, the result of the merger being a recent explosion of what I propose to call *slow lit*. This label refers intentionally to chick lit, this "strand within women's genres which explores in a humorous manner the social and emotional lives of its protagonists"[54] or, in the words of Susan Ferriss and Mallory Young, "a form of woman's fiction on the basis of subject matter, character, audience, and narrative style."[55] The category has been distinguished not on the basis of a type of protagonist, as in the case of other popular subgenres of chick lit, such as, for example, "nanny lit" or "lad lit," but on the basis of a certain style and structure of narration which embraces popular novels whose plots are constructed around the idea of urban dwellers who at

---

51 Parkins, "At Home in Tuscany," p. 270.
52 Parkins, "At Home in Tuscany," p. 258.
53 Jenny Shaw, "'Winning Territory.' Changing Place to Change Pace," in *Timespace. Geographies of Temporality*, eds. Jon May, Nigel Thrift (London: Routledge, 2001), p. 120.
54 Katarzyna Smyczyńska, "'The Curse of the Pink Cover': Chick Lit in Chick Culture," in *Beyond 2000. The Recent Novel in English*, ed. Ewa Rychter (Wałbrzych: Wydawnictwo Państwowej Wyższej Szkoły Zawodowej, 2011), p. 15.
55 Susan Ferriss, Mallory Young, "Introduction," in *Chick Lit. The New Women's Fiction*, eds. Susan Ferriss, Mallory Young (New York: Routlege, 2006), p. 3.

some point of their lives decide to liberate themselves from the tyranny of the accelerated urban lifestyle and move to the country, where they learn how to "live in the present in a meaningful, sustainable, thoughtful *and pleasurable* way" [emphasis original],[56] to cite once again Parkins and Craig's oft-quoted description of the nature of slow living.

One of the most characteristic features of chick lit is a narrative style which allows readers to identify with the protagonist. The confessional modes of diaries, letters, emails and first person narration are used to "craft the impression that the protagonist is speaking directly to readers."[57] The air of authenticity, which is "frequently missing from women's fiction of the past,"[58] makes it possible for readers to identify with literary characters. By creating humorous and often ironic but usually realistic descriptions of everyday life lived by heroines, who can be "rude, shallow, overly compulsive, neurotic, insecure, bold, ambitious, witty, or surprisingly all of the above,"[59] chick lit provides readers with an insight into the lives of women who are putatively like them and who have to confront problems and obstacles they know from their own lives. Ferriss and Young emphasize that many chick lit stories have been inspired by novelists' own experiences (Jennifer Weiner's "rough breakup," or Emma McLaughlin and Nicola Kraus's professional experience which led them to write *The Nanny Diaries*), the result being "the perception that chic lit is not fiction at all."[60] Katarzyna Smyczyńska notes:

> British chick lit continues to rework contemporary social anxieties. The main themes that the novels explore are extremely diverse and include career, family problems, intimate relationships, pregnancy and parenthood, friendship, diet, travelling, romance, and shopping. The pleasure of reading probably consists in the fact that each character is a contemporary 'everywoman' with whom it is easy to identify [...].[61]

When viewed as a product of a specific cultural moment, whose character reflects the anxieties and contradictions of the present, chick lit turns out to highlight dilemmas of women (and men) living in the conditions of a globalized consumer culture, which constantly exerts its influence on their lives.

---

56   Parkins, Craig, *Slow Living*, p. ix.
57   Ferriss, Young, *Chic Lit. The New Women's Fiction*, p. 4.
58   Jennifer Weiner, quoted by Ferris, Young in *Chick Lit. The New Women's Fiction*, p. 4.
59   Website "chick-lit.us," quoted by Ferris, Young in *Chick Lit. The New Women's Fiction*, p. 4.
60   Ferriss, Young, *Chick Lit. The New Women's Fiction*, p. 4.
61   Smyczyńska, "'The Curse of the Pink Cover,'" p. 19.

Chick lit, this "commercial tsunami,"[62] has produced a number of subgenres, such as "mom lit," "sistah lit," "nanny lit," "Christian lit," "widow lit" and "lad lit." Considering the popularity of the genre, the marketing decision to extend the thematic scope of chick lit and embrace elements of the increasingly popular social trend of slowing down was in a sense predictable. I propose to include in the category of slow lit humorous novels written mainly by women, for women audiences, in which a protagonist, usually a successful and fashionable professional in her thirties, alone if she is single, or with her family, decides to make a new start somewhere outside the city. British novels that fall into this category include Tessa Hainsworth's autobiographical Cornish trilogy: *Up with the Larks: Starting Again in Cornwall* (2009), *Seagulls in the Attic. Making a New Life in Cornwall* (2010), and *Home to Roost: Putting Down Roots in Cornwall* (2012), Karen Wheeler's *Tout Sweet: Hanging Up My High Heels for a New Life in France* (2009) and *Toute Allure: Falling in Love in Rural France* (2010), a pair of novels about a former fashion editor, the owner of a fashionable flat in London, who settles in Western France, and Daniel Butler and Bel Creve's *Urban Dreams, Rural Realities. In Pursuit of the Good Life* (1999), the story of a couple's relocation from London to Wales, told in contrasting voices by both of them.

It is tempting to view slow lit as a part of an energetically developing branch of popular commercial culture which has recognized the marketing potential of a growing sense of dissatisfaction with consumerist ideologies. The branch caters for the needs of thousands of people who dream of moving from city to the country by providing them with ideas and useful self-help books which seem to show how to achieve these dreams. While television documentaries like *River Cottage* offer insights into the everyday life of a downshifter, popular guides offer advice to all those who are considering a major change in their lives. *Out of Your Townie Mind. The Reality behind the Dream of Country Living* by Richard Craze is praised by the guru of British downshifting, the author of the *River Cottage* documentary series, Hugh Fearnley-Whittingstall, for its realistic approach when he says that "Richard Craze yanks the rose-tinted spectacles from the rural idyll and tramples them in the mud" and describes the book as "a kind of Feel-the-Fear-but-Do-It-Anyway for wannabe downshifters."[63] *The Downshifter's Guide to Relocation. Escape to a Simpler, Less Stressful Life* by Chris and Gillean Sangster helps prepare anyone wishing to settle in the country by discussing "the

---

62 Kate Zernike, quoted by Ferriss, Young in *Chick Lit. The New Women's Fiction*, p. 2.
63 Richard Craze, *Out of Your Townie Mind. The Reality behind the Dream of Country Living* (Great Ambrook: White Adder Press, 2004), front cover.

pros and cons of living in different parts of the country,"[64] while *Downshift to the Good Life* by Lynn Huggins-Cooper gives "52 brilliant ideas"[65] to make relocation work. Slow lit may be read as a literary counterpart of popular guidebooks to relocation and downshifting.

If we agree that chick lit may be approached from the perspective of "its engagement with contemporary culture,"[66] it may sound reasonable to view slow lit in a broader context of changing cultural landscape. Seen as a part of a wide social and cultural trend, slow lit represents some of the dilemmas of urban dwellers, their quest for meaning in life and their refusal to be swamped by the flood of an accelerated urban life. Tessa Hainsworth's *Up with the Larks: Starting Again in Cornwall* may serve as an illustration of some of the most distinctive features of British popular relocation novels. The way she describes living in a Cornish village is marked by a departure from the convention of what often sounds like a Tuscan utopian dream. What she proposes instead, in a manner that recalls the warning tone of some of the guides to relocation, is to put aside the rose-tinted spectacles and confront the harsh realities of the relocation project, which nevertheless has to have a happy ending, and which is represented in a manner characteristic of chick lit.

Hainsworth's Cornish trilogy is based on the author's personal experience, which she uses to create a semi-autobiographical story about her family leaving London and settling down in a Cornish seaside town. With its focus on a "specific age, race and class: young, white, and middle,"[67] chick lit offers a portrait of a contemporary woman living her life in an urban environment. Hainsworth is, to some extent, a typical heroine of chick lit. Young, white, middle-class, although not single (she has a husband and two children) like many other chick lit heroines, she is a professional woman working as UK marketing manager for a famous international cosmetics company, The Body Shop, who at a certain point in her life discovers that she has become enslaved by the life she used to like so much. This is how she describes her everyday life in London:

> The company had expanded far beyond anyone's expectation, and in the past few years I'd found it more stressful than inspiring. Though I was often in London where most of my job was based, we lived in a faceless commuter neighbourhood on the outskirts of the city in a house that often seemed less a home than a hotel where I crashed out after meetings, conferences and business entertaining. When I wasn't travelling, I came home

---

64   Chris and Gillean Sangster, *The Downshifter's Guide to Relocation*, p. 2.
65   Lynn Huggins-Cooper, *Downshift to the Good Life* (Oxford: Infinite Ideas Limited, 2004).
66   Ferriss, Young, *Chick Lit. The New Women's Fiction*, p. 3.
67   Ferriss, Young, *Chick Lit. The New Women's Fiction*, p. 8.

some nights well after the kids were in bed, shattered after the day's commuting. Getting to and from work was a nightmare of crowded trains that arrived late, tube lines closed, buses missed and taxis that never came. I hardly saw my kids, hardly saw Ben. I went from one extreme to the other, either revved up from the stresses of the job, or totally limp and exhausted.[68]

This description of such a hectic life, which is physically and emotionally exhausting, has the potential to appeal to thousands of professional women working in cities, dreaming about a change, and wishing to liberate themselves from the tyranny of an accelerated culture. Regardless of the fact that not every woman can be a top manager for the The Body Shop, or work for a fashion magazine as in the case of Kate Wheeler (another popular slow lit author), identification with characters takes place beyond job differences because the sense of disillusionment with the promises of urban consumer lifestyle is what many working women in cities have in common. Chick lit's embeddedness in daily life is the genre's most unmistakable trait. Ferriss and Young quote a popular American web site, www.chicklit.us, which asserts that the readers "want to see their own lives in all the messy detail, reflected in fiction today."[69]

Slow lit is permeated by a sense of displeasure with consumerism. Being a part of the rat race ("I've got about six people after my job as it is" (*UwtL*, 6) coupled with the indifference generated by the policy of profit maximization ("I'd had to fire a single mum against my will [...]" (*UwtL*, 7) produces a sense of alienation and separation from the character's professional life. Tessa says to her husband: "[y]ou know how the company's grown from the small cosy firm I started with to this huge multinational. I don't speak its language any more" (*UwtL*, 9). For Karen Wheeler an unwillingness to participate in sustaining the mythology of the world of fashion turned out to be a driving force which made her abandon her posh life in London:

> I was tired of conspiring in key fashion myths: that it's necessary to spend £600-plus on a new designer handbag every six months, or that a grown-up woman could look good in a ra-ra skirt, micro-shorts or whatever unseemly trend designers were pushing that season. I also felt guilty about persuading readers to rush out and buy 'must-have' items that I knew were 'must-nots' and that would end up on a fast track to landfill within six months.[70]

---

68   Tessa Hainsworth, *Up with the Larks. Starting Again in Cornwall* (London: Preface Publishing, 2009), p. 5. Throughout the chapter this book will be referred to as *UwtL*.
69   Feriss, Young, *Chick Lit. The New Women's Fiction*, p. 3.
70   Karen Wheeler, *Tout Sweet: Hanging Up my High Heels for a New Life in France* (Chichester: Summersdale, 2008), pp. 8–9.

While chick lit is often criticized for its "investment" in the world of "fashion and cocktails," its defenders like to see it as a "reflection of consumer culture."[71] In slow lit the professional world is no longer "the ultimate chick challenge"[72] since the protagonists have succeeded and made their careers. In slow lit the challenge lies somewhere else. It has got its roots in a critical examination of the side-effects of the consumer lifestyle and the model of the good life it offers. The relocation motif, which is used in slow lit as a narrative axis, helps promote "the attractions of a post-consumerist life-style"[73] and exemplifies the possibility of a successful retreat and liberation from the world of acceleration and excess. Both Tessa Hainsworth and Kate Wheeler are professional women whose experiences can be read as representative of "some emerging forms of self-interested disaffection with consumerism on the part of affluent consumers themselves."[74]

The sense of humour, which marks the genre's tone, often stems from the clash between a character's former and present life. Karen Wheeler confesses: "I don't even have clothes for this kind of life. After a decade and a half of working in fashion, most of my wardrobe is designed for going to cocktail parties – or, at the very least, breakfasts at Claridges – and my shoes are so high I need a Sherpa and an oxygen tank to wear them."[75] Ferriss and Young observe that chick lit's main concern is with the world of beauty and its pillars: "shopping, fashion and consumerism," which leads to "an arguably obsessive focus on skin-deep beauty."[76] Seen in this light, slow lit seems to be a form of *post*-chick lit, with its portraits of women who are no longer fashion slaves, and who leave behind the world of urban glamour and instead, like Tessa Hainsworth, are ready to be grateful for the unflattering, waterproof jackets the workers of the Royal Mail, one of whom she has become, are obliged to wear.

The main difference between the past and present is humorously illustrated by the opening of Hainsworth's *Up with the Larks*. In the first chapter, entitled "November," we read:

> My first day as a postwoman for the Royal Mail and the first shock is odour. Male sweat – the sorting room reeks of it – not rancid or unclean, just ordinary bloke secretions, with all that testosterone crammed together in one room. I feel as though I'm in

---

71 Ferriss, Young, *Chick Lit. The New Women's Fiction*, p. 4.
72 Ferriss, Young, *Chick Lit. The New Women's Fiction*, p. 7.
73 Soper, "The Other Pleasures of Post-Consumerism," p. 31.
74 Soper, "Alternative Hedonism…," p. 567.
75 Wheeler, *Tout Sweet*, p. 6.
76 Ferriss, Young, *Chick Lit. The New Women's Fiction*, p. 11.

the wrong story, the one about warriors and adventures and manly prowess; a story that begins far too early, before dawn (*UwtL*, 1).

The transition from Hainsworth's professional world of "the sweetest perfumes, lemongrass and lavender, mimosa and magnolia, fruity and flowery fragrances" (*UwtL*, 2) to the world of sweaty odours is a marker of the clash between expectations and reality and foretells the challenges of a new life. It is illustrative of the author's intention not to multiply idyllic representations of life in the country; what she proposes instead is a story of overcoming difficulties and struggling with the reality of country life in Cornwall which, although rewarding in many respects, still requires from her a great deal of determination, at least in the beginning.

"We were so well-prepared for our new Cornish life that the reality of it, when it hit, was doubly hard to come to terms with" (*UwtL*, 17), confesses Hainsworth, and she has to admit that "[…] moving to Cornwall has not been the idyllic move" (*UwtL*, 24) they had hoped it would be. Armed with a precise master plan for their paint-your-own-pottery business and the money they had after selling their London property, the family starts its life in Cornwall to find out that the price of houses has gone up, finding a job is extremely difficult and the money they have is not going to last long. They move into a house for which they have to pay more then they planned and which requires serious renovation, and Tessa is thrilled when finally, after many failures to find any job, she has a chance to become a postwoman.

"'The rural' is of course often subject to romantic and nostalgic valorizations,"[77] Parkins and Craig claim, and while this statement clearly applies to a majority of Tuscan novels, slow lit seems to be constructed along slightly different lines. It seems to be less utopian because it plays out the element of contrast between the dream and reality, which on the one hand is a source of humour, and on the other shows the changing cultural and economic landscape of the countryside and shows how the urban/rural divide has recently been complicated by global contexts.

A realistic approach to the representation of Cornish reality allows Hainsworth to register a vital cultural moment brought about by urban deconcentration. Her observations on the Cornish housing market and its change under the influence of a flow of wealthy, urban second-homers record a significant change southern Cornwall has recently undergone. Parkins and Craig note that the rural is "becoming a much more complex space, and its transformation

---

77 Parkins, Craig, *Slow Living*, p. 84.

through engagement with other places and global flows provides opportunities as well as threats [...]."⁷⁸ Cornwall's transformation and its engagement with "global flows" is marked by the fact that due to a recent explosion of interest in rural Cornwall, the influx of second-homers has reconfigured the Cornish landscape and sent the price of properties through the roof. Hainsworth's narrative offers an insight into the mechanisms of this reconfiguration when she describes one of the posh streets of her town, which the locals call Millionaire's Row:

> The houses are huge and opulent, all facing the sea but with massive lawns and gardens on the slopes in front so that they cannot be seen by the likes of us commoners. The locals have nothing against millionaires, it's just that many of these houses are empty, holiday homes for the extremely wealthy, sometimes used no more then one or two weeks a year. Not long ago a helicopter circled one of them at the very edge of the sea, set in a secluded, private woodland. The place has just come onto the market. There was one passenger in the private helicopter. He took a quick aerial view and bought the property. There were rumors that he was a pop star or even royalty, for we have that here too, but it turned out he was yet another businessman looking for a second home to buy with his Christmas bonus (*UwtL*, 48).

While it is true to say that due to global changes "[r]ural populations have extended their networks, widening their social space and economic scope,"⁷⁹ the consequences these changes have brought about are not only positive. As Hainsworth notes, it is very often the case that the locals "can't afford to buy even a modest home in the country where they grew up" (*UwtL*, 48) which leads to a migration of professional people from Cornwall. "[T]reated as just another disposable commodity,"⁸⁰ a house loses its capacity to display its "cultural and emotional power":

> In modern western societies, the house owes its cultural and emotional power to its capacity to separate itself ideologically from the public spaces of everyday life – what Marc Augé calls the 'non-places' such as motorways, subways, commuting trains and office parks, which encourage functional, transient behavior and produce a peculiar mix of alienation and liberating self-erasure (1995). The extent to which the house has come to be seen as a refuge from the non-place requires a great deal of symbolic work to conceal the sameness of houses, and their connection to these collective routines and temporary communities.⁸¹

---

78   Parkins, Craig, *Slow Living*, p. 84.
79   Jacinthe Bessière, quoted by Parkins, Craig in *Slow Living*, p. 84.
80   Joe Moran, "Housing, Memory and Everyday Life in Contemporary Britain," *Cultural* Studies, Vol. 18, No. 4 (2004), p. 621.
81   Moran, "Housing, Memory and Everyday Life in Contemporary Britain," p. 608.

In the light of the above we see that what Hainsworth describes is *a* house which is deprived of it capacity to become *the* house and which is still an element of the space of flow rather than that of settlement. Neither a sense of community nor a sense of connection with a specific area can start to grow in conditions which make it impossible for a house owner to establish any sort of meaningful relation with the local area and its inhabitants, "who coexist in the same territory and who collectively give their territory identity and value."[82]

Difficult as it might be, the life Hainsworth describes compensates for the sense of financial insecurity and everyday hardships that an ex-London-top-manager-learning-to-be-a-postwoman-in-a-Cornish-town has to confront. Liberated from the tyranny of fast time, this "fragmented and rushed temporality,"[83] she starts a passage from the lost to the recovered self. The sense of imbalance within her, which has brought about the feelings of loss and anxiety, is gradually replaced by a sense of harmony and reconnection to her real self. The inner journey towards her real self which she starts in Cornwall has a symbolic beginning. During one of the family's trips to Cornwall, which they often made before selling their London house, Hainsworth goes for a walk on the beach where her body opens up to the multisensory experience:

> My senses were being bombarded: the earthy smells of sea and stone and damp, the sounds of waves churning over the pebbly beach and of sea birds calling to each other overhead, and I could almost taste the salt in the air, it was strong and pungent (*UwtL*, 12).

What she experiences at that moment is a form of awakening and reconnection with the world of nature from which she has been separated. It is the suppressed wisdom of her "unexercised, unhealthy and exhausted body" (*UwtL*, 5) that overwhelms her and helps her to crystallize what she had sensed but was not ready to acknowledge: "[t]*his is where we must go. This is where we belong, by the sea, in this place*"[emphasis original] (*UwtL*, 12). She feels that despite the low spring temperatures something is telling her to go for a swim in the ocean, which turns out to be a spiritually purifying experience:

> The knowledge, the certainty of my feelings made me suddenly wild and exhilarated. […] I didn't hesitate. I wanted now to *feel* [emphasis original] the sea on my body, I wanted to actually taste the salt water on my lips. I wanted a baptism, too, although I didn't form that thought till later. I wanted to immerse myself in my new certainty (*UwtL*, 12).

---

82 Parkins, Craig, *Slow Living*, p. 85.
83 Eriksen, *Tyranny of the Moment*, p. 148.

A swim in the ocean, "a baptism," is a symbolic beginning of a new life, a rite of passage from the world of excess, speed and imbalance to the world which permits rediscovery of the self. When she finishes her swim and reaches the shore, she feels she is now capable of seeing the world and not just looking at it:

> The mist had gathered again as I staggered out, feeling like the first creature to crawl on dry land, looking around me at the awesome world I had not truly looked at before. I hadn't had the *time* [emphasis original] to look before, or, if I *had* a rare moment to myself, the whirling voices in my head – planning, worrying – kept me from seeing anything (*UwtL*, 13).

Her immersion in the natural environment and direct, sensually experienced reality provides her with, in the words of Abram, "the sole solid touchstone for an experiential world now inundated with electronically-generated vistas and engineered pleasures."[84] It is this "return to things themselves,"[85] in Maurice Merleau-Ponty's famous words, that is the starting point for Hainsworth, a point of departure to reclaim herself. The appeal of the life Cornwall offers her consists in a chance to approach everyday life "with care and attention,"[86] which Parkins and Craig see as defining qualities of slow living:

> I look out over the damp green fields where a few sheep are idly grazing. I'm not thinking of anything, which happens increasingly since we've moved her. In London, my mind was always racing, always one step ahead of the present, thinking of things that needed doing or planning, figuring out how to order the future instead of concentrating on now.
>
> Here I'm beginning to live in the moment and it's bringing me a peace and happiness far beyond anything I've known before.[…] Now I feel as if I'm *living* [emphasis original] life instead of hassling about it (*UwtL*, 146).

Hainsworth's novel is an an account of her attempt to re-make her life and create solid foundations of the reclaimed self. The fact that she is now able to reconnect with herself results in her opening up to the world, both human and non-human. And her immersion in the present has not only an individual but also a social dimension. Reading Alberto Meluci's analysis of the significance of inner time, Parkins and Craig note that "the cultivation of attention to the present has positive potential beyond the individual subject to mobilize new forms of political

---

84 Abram, *The Spell of the Sensuous*, p. x.
85 Maurice Merleau-Ponty, *Phenomenology of Perception*, trans. Colin Smith (London: Routledge & Kegan Paul, 1962), p. ix.
86 Parkins, Craig, *Slow Living*, p. ix.

investment and revivify everyday life."[87] Hainsworth newly acquired ability to live in the present brings about such a mobilization of energy that results in the acts of reciprocity and community-oriented thinking as "her deep individual experience transforms itself into a social energy for change."[88] Her work for the community not only lessens the sense of aloofness she often experiences, but also helps her see herself in the context of a community, a member of which she finally becomes

Conclusions

The recent proliferation of narratives about downshifting in the country is part of a wider context of continuity and change. Popular narratives of slowness are the product of a merger: a literary traditon of representing Italy as an idealized place to experience spiritual awakening and refashion one's self, and chick lit, with its humorous depictions of lives of contemporary, middle-class, professional women, have produced their "slow" versions which by emphasising the country and city divide popularize a slow living philosophy and the lifestyles it promotes.

The history of the country and city divide, which forms an organising principle around which plots in the "Tuscan farmhouse" literature and slow lit novels are constructed, has roots in the classical imaginary. From the very beginning it has been invested with, as Raymond Williams puts it, "powerful feelings":

> On the country has gathered the idea of a natural way of life: of peace, innocence, and simple virtue. On the city has gathered the idea of an achieved centre: of learning, communication, light. Powerful hostile associations have also developed: on the city as a place of noise, worldliness and ambition; on the country as a place of backwardness, ignorance, limitation.[89]

What popular narratives of slowness have in common is that they have crystallized the divide by adopting a tone of idealization (even when they attempt to describe realistically the hardships of living in the country as we have seen in the case of Hainsworth's novel) when talking about the country, and a tone of critical evaluation when talking about the city. The representation of the country and the city that the novels provide fixes their images, the result being a somehow formulaic evolution of the plot in which characters experience the passage from

---

87 Parkins and Craig, *Slow Living*, p. 5.
88 Alberto Melucci, quoted by Parkins, Craig in *Slow Living*, p. 5.
89 Raymond Williams, *The Country and the City* (London: The Hogarth Press, 1993), p. 1.

a world of chaos and excess (the city) to a world of harmony and balance (the country).

"[P]eace, innocence and simple value," which are positively associated with living in the country, resonate with the mindfulness, balance and sustainability that slow living promotes as the values of alternative lifestyles which are deeply rooted in a desire to resist the tyranny of the moment and consumerist and work-dominated modes of living. The country is represented as a space where a "mindful use of time"[90] helps one rediscover one's real self, find pleasure in alternative practices and enter meaningful relations with other people and with the natural environment. Numerous representations of a multisensory experience of nature, which the novels abound with, are illustrative of the departure from the paradigm of visual experience promoted by contemporary culture. The turn towards an extension of the modes of perception corresponds with a quest for "the heightened awareness"[91] that slow living advocates.

The popularity of relocation novels (and other relocation "products" such as television programmes and guidebooks) may be explained from the perspective of a sense of ambivalence about the advantages of an accelerated urban lifestyle promoted by consumerist global culture. The premise of alternative hedonism that "consumerist lifestyles do not have to be replaced by 'stone-age simplicity' and puritan self-denial" and that "other ways of living and forms of pleasure can and must be developed"[92] is well illustrated by stories about downshifters who experience the passage from a world of clutter to a world of clarity by leaving behind the urban rush and settling in the country. The appeal of the relocation narratives seems largely to consist in the promise that relocation does not have to be an unfulfilled dream but may become a strategy for resistance against consumerism and its displeasures.

---

90   Parkins, Craig, *Slow Living*, p. 4.
91   Parkins, Craig, *Slow Living*, p. ix.
92   Thomas, "Alternative Realities…," p. 681.

Conclusions: Post-slow? Between Strategies and Props

> Slow is the new fast, and there's a growing
> Movement-with-a-capital-M to prove it.
> Preston Lerner, "Slow cars – the Joy of Slow"

> Let's embrace speed, for it represents the marvelous
> of the modern era. But let's always check our brakes.
> Paul Morand, *On Speed*

In 1996 Ivan Illich declared that "[t]he Age of Speed had a beginning, and we talk about its history because we witness its end."[1] The surprising announcement of the end of the age of speed calls for a critical re-examination of speed and generates a series of questions, some of which will probably for some time remain unanswered. Provided that the statement was not only made to provoke a reaction from those who try to outrun themselves, it may imply that regardless of how speedy we may become in the future, we have already reached a point after which further acceleration is unlikely to define social relations in the way it has been defining them over the past two centuries. We may assume that further increase in the speed of life will bring faster communication and transportation and quicker delivery of pleasures, but the quality and character of the changes brought about by further technological advances will neither restructure existing social systems nor generate radically new models of social interaction. Thus, to proclaim the end of the age of speed is to assert that the prospects of new speeds are not likely to create new visions of the world and new reconceptualizations of time and space. Neither are they likely to inspire enthusiasm similar to that which was experienced by the speedsters at the beginning of the 20th century. Jeffrey T. Schnapp may be right when he says that "the romance of speed might well seem to belong to the past."[2]

Illich's claim may also suggest that a sense of the end of an epoch marked by an obsession with speed, record-breaking and overcoming the limits of human

---

1 Ivan Illich, Matthias Rieger, Sebastian Trapp, "Speed? What Speed?" in *Speed – Visions of an Accelerated Age*, eds. Jeremy Millar, Michiel Schwartz (London: The Photographer's Gallery and the Trustees of the Whitechapel Art Gallery, 1998), p. 154.
2 Jeffrey T. Schnapp, "Fast (Slow) Modern," in *Speed Limits*, ed. Jeffrey T. Schnapp (Milan: Skira, 2009), p. 35.

physicality can also be analyzed in terms of a growing sense of popular dissatisfaction with accelerated everyday life and a rising awareness of the often disastrous side-effects of technological acceleration. The end of the age of speed may be approaching as we openly declare the need to slow down, replacing praise of speed with praise of slowness. Though unlikely to challenge the popularity of speed, support for deceleration may be interpreted as a manifestation of a crisis in the belief that speed is a guarantor of success and wellbeing. Cathy Left summarizes this concern, declaring that "we are at a moment in which understanding speed and its limits matters more than ever."[3]

As the preceding chapters have shown, discourses of the slow now permeate a great variety of fields, philosophies and representational practices, from the discourse of food culture and new territoriality to slow travel and slow lit. Naturally, if we take into account the modernist/industrial origin of the phenomenon and its subsequent embodiments in a redefined notion of pleasure, in the critique of the culture of immediacy and in the new concept of hedonism, we will notice that today's forms of cultural deceleration include them all, although in different proportions and through different interpretations. Somehow paradoxically, the culture of the slow is a culture on the move, always-already evolving in accordance with increased critical awareness of its practitioners and hence successfully escaping ultimate definitions and univocal classifications. Its popularity lies in its widespread potential to generate intellectual alternatives, ideological counterparts and less conventional life-styles, which all constitute a powerful counter-ideology of social existence, an ideology one of whose major strengths is the fact that it is at the same time conservative and progressive.

Even though to characterize slow culture as conservative may seem odd at first sight, upon closer look it does not appear so unacceptable. Slow culture proposes a system of integrated practices and ideologies which together form a philosophy of life, a relatively stable counterbalance to visions of the critical condition of humanity, post-modern fragmentation, instability and exhaustion that contemporary cultural debates constantly emphasize. The discourse which slow culture has created is holistic, i.e. it offers a chance to embrace almost all spheres of life, from work to consumption to travelling and leisure, and these can therefore create a coherent lifestyle. In other words, though it was originally inspired by a critical reaction to the culture of immediacy, slow culture reaches beyond mere negation of speed as an operating principle and offers a whole range of alternative, independent and philosophically autonomous paradigms which find

---

3   Cathy Left, "Preface," in *Speed Limits*, ed. Jeffrey T. Schnapp (Milan: Skira, 2009), p. 14.

their embodiment both in self-regulating theories and in self-determining social practices.

Still, the possibility of creating an all-embracing and coherent lifestyle philosophy which may be enacted in everyday life only partially explains the recent popularity of slow living. Another possible explanation is in the progressive shift of accent which slow living offers, suggesting the opportunity for change in an age of "political disengagement, whereby citizens have become apathetic, isolated form the political process and alienated from the values of citizenship,"[4] to quote Geoff Andrews, author of *The Slow Food Story. Politics and Pleasure*. The complexity of life in the modern world is hardly conducive to building solid and stable roads which would guarantee safe conduct through the hazards of the contemporary landscape. Scholars seem to agree in their evaluation of the character of modern life and although they use different concepts and metaphors, what resonates in their works is a sense of global crisis which appears to affect social, political, economic and cultural dimensions and structures of the present. Scott Lash talks about risk culture,[5] Manuel Castells sees "a world of uncontrolled, confusing change"[6] and Tony Judt diagnoses that "[i]ll fares the land."[7] While for Anthony Giddens the world of high modernity is "apocalyptic, not because it is inevitably heading towards calamity, but because it introduces risks which previous generations have not had to face,"[8] Ulrich Beck registers "the horrific panorama of a self-endangering civilization"[9] marked by the rise of risk society which is "a *catastrophic* society [emphasis original]."[10] A wide range of phenomena contribute to the sense of individual, communal and national instability: overdependence on technology, whose infallibility has been falsely taken for granted, cyberwarfare and the threat of computer viruses, unsustainable consumption of natural resources, climate change, globalization,

---

4   Geoff Andrews, *The Slow Food Story. Politics and Pleasure* (Montreal: McGill-Queen's University Press, 2008), p. 86.
5   See: Scott Lash, "Risk Culture," in *The Risk Society and Beyond. Critical Issues for Social Theory*, eds. Barbara Adam, Ulrich Beck, Joost Van Loon (London: Sage, 2005) and Ulrich Beck, *Risk Society* (London: Sage, 1992).
6   Manuel Castells, *The Rise of the Network Society* (Malden, MA.: Wiley-Blackwell, 2010), p. 3.
7   See: Tony Judt, *Ill Fares the Land* (London: Penguin, 2011).
8   Anthony Giddens, *Modernity and Self-Identity. Self and Society in the Late Modern Age* (Stanford: Stanford University Press, 1991), p. 4.
9   Ulrich Beck, *Risk Society: Towards a New Modernity*, trans. Mark Ritter (London: Sage, 2000), p. 10.
10 Beck, *Risk Society*, p. 24.

international terrorism, cloning, genetic modification of food, economic crises and instability of financial markets are just a few on a long list of contemporary threats. In light of the above, the discourse of slow life and the images it creates turn out to provide a sense of progressive counterbalance to the imagery of social, economic and ecological apocalypse. Slow life may be interpreted as suggesting a possibility of the good life, which is represented not so much as a utopian dream, but as a space of meaningful changes available to everyone. Apart from the ideological support, it offers practical suggestions which may be realized in everyday life and on a daily basis, both individually and collectively. The rhetoric of slow life is the rhetoric of agency and empowerment.

Slow life started as a reaction to the dominant ideologies of progress and development discreetly supported by ideals of efficiency and productivity. As a response to a sense of cultural exhaustion, the discourse of slowness not only highlights a sense of individual and collective power to change the existing social structures but also creates its own mythology which helps anchor the ideals of slow life in the context of tensions and anxieties brought about by the accelerated and globalized world. Popular slow narratives are emblematic of a desire to remap contemporary culture and identify zones of danger as well as zones of retreat. The new cartographies of happiness are moulded by increasingly popular claims that it is simplicity rather than excess that may activate new sources of life satisfaction. Instances of romanticization and idealization of slow life are illustrative of a sense of nostalgy for the idealized states of culture, which, as Wojciech Burszta and Waldemar Kuligowski observed, is a response to the sense of insecurity, uprootedness and acceleration.[11] Burszta and Kuligowski also suggest that in the light of the changes brought about by the rise of the information society the questions of developmental continuity and transformations become more significant than ever. As they imply, continuity exists but at the same time it is broken and violently disrupted by the rise of the practices and routines which irreversibly change the nature of social, cultural and economic relations. The discourse of slow culture articulates the significance of a reconnection to people and places and expresses it in the language of affirmation and reconstruction.

As has been noted in the Introduction, to study the culture of the slow at this particular moment seems viable for a number of reasons among which we have distinguished the fact that in the last few decades the movements gathered under the headline of slowing down have ceased to be a novel attraction and have

---

11 Wojciech Józef Burszta, Waldemar Kuligowski, *Sequel. Dalsze przygody kultury w globalnym świecie* (Warszawa: Muza S. A., 2005), pp. 44–56.

evolved into one of the defining traits of contemporary cultural practices instead. It is precisely for this reason that we can think about slow culture in terms of its evolutionary stages, which are marked, respectively, by resistive origins, maturation and development and, in the last stage, by the gradual shift of the slow from cultural peripheries to cultural mainstream, often at the price of its commodification and commercialization. The *post-slow* stage, as I propose to call the present moment, does not herald its end; rather it points to certain processes which are responsible for the generation of sideline practices which adopt slowness not so much for its resistive potential but rather as a prop of a trendy lifestyle. Suspended between practical strategies of resistance and often romanticized dreams and mythologies of the good life, slow living is not free from inner tensions and often displays its contradictory character. Nick Osbaldiston rightly observed that "the slow adjective is broadened to the point where it becomes an empty signifier, filled with endless an ultimately meaningless interpretations."[12] No longer a strategy of resistance or mode of expression of disillusionment with the quality of contemporary life, slow products and practices have become props of a certain type of post-modern/urban identity whose premises are defined largely in terms of lifestyle, recognizable and fashionable. This fact, however, should neither be overestimated nor demonized as an illustration of an ultimate failure of the slow clause. Rather, it indicates the inevitability of certain processes which take place in the culture marked by the plurality of choices. Regardless of what may be interpreted as the unquestionable appeal of slow living, it is hard to expect that a vision of slow, harmonious and sustainable life will make everybody abandon or modify their everyday routines. This is because the contemporary flea-market of attractive lifestyles sometimes evolves into one of in-depth identity-forming philosophies and since the border between the former and the latter is far from being a clear-cut one, the culture of the slow, whether appreciated or ridiculed, is and will remain one of contemporary culture's most attractive offers, in equal measures trivial and complex.

---

12   Nick Osbaldiston, "Conclusion: Departing Notes on the Slow Narrative," in *Culture of the Slow. Social Deceleration in an Accelerated World*, ed. Nick Osbaldiston (London: Palgrave Macmillan, 2013), p. 179.

Bibliography

Abram, David. 1997. *The Spell of the Sensuous*. New York: Vintage.

Andrews, Cecile. 2006. *The Slow is Beautiful. New Visions of Community, Leisure and Joie de Vivre*. Gabriola Island: New Society Publishers.

Andrews, Geoff. 2008. *The Slow Food Story: Politics and Pleasure*. Montreal: McGill Queens University Press.

Archer, William. 1903. "Henley's A Song of Speed." *The New York Times*, 20 June. Web: http://query.nytimes.com/mem/archivefree/pdf?res=F20F14F73D5D11738DDDA90A94DE405B838CF1D3. Accessed 4 March 2014.

Augé, Marc. 1992. *Non-Places. Introduction to an Anthropology of Supermodernity*. Trans. John Howe. London: Verso.

Barton, Robin, Hayley Cull. 2010. *Slow London*. Melbourne: Affirm Press/Hardie Grant Books.

Basista, Andrzej. 2004. "Spojrzenie z ukosa. Tęsknota za ulicami i placami." *Autoportret. Pismo o dobrej przestrzeni*, 1[6]. Kraków: Małopolski Instytut Kultury.

Bauman, Zygmunt. 2010. *44 Letters from the Liquid Modern World*. Cambridge: Polity Press.

Beck, Ulrich. 2000 [1992]. *Risk Society: Towards a New Modernity*. Trans. Mark Ritter. London: Sage.

Beerbohm, Max. 1946. *Mainly on the Air*. London: William Heinemann Ltd.

Berardi, Franco. "Futurism and the Reversal of the Future." Web: http://www.generation-online.org/p/fp_bifo8.htm. Accessed 20 July 2011.

Bigell, Werner. "Distinction but not Separation: Edward Abbey's Conceptualization of Nature." *A Dissertation for the Degree of Doctor Philosophiae*, University Of Tromsø, Faculty of Humanities, Department of English, August 2006. Online dissertation: http://munin.uit.no/bitstream/handle/10037/1227/thesis.pdf?sequence=1. Accessed 30 May 2012.

Bois, Yve-Alain. 2009. "Slow (Fast) Modern," in *Speed Limits*. Ed. Jeffrey T. Schnapp. Milan: Skira.

Bourdieu, Pierre. 1984. *Distinction: A Social Critique of the Judgment of Taste*. Trans. Richard Nice. London: Routledge.

Buczyńska-Garewicz, Hanna. 2006. *Miejsca, strony, okolica. Przyczynek do fenomenologii przestrzeni*. Kraków: Universitas.

Burszta, Wojciech Józef, Waldemar Kuligowski. 2005. *Sequel. Dalsze przygody kultury w globalnym świecie*. Warszawa: Muza S.A.

Campbell, Colin. 1987. *The Romantic Ethic and the Spirit of Modern Consumerism*. Oxford: Blackwell.

Castells, Manuel. 2010 [1996]. *The Rise of the Network Society*. Malden, MA.: Wiley-Blackwell.

Castells, Manuel. 2010 [1997]. *The Power of Identity*. Malden, MA.: Wiley-Blackwell.

Chambers, Robert, ed. 1832. *The Book of Days: A Miscellany of Popular Antiquities in Connection with the Calendar*. Vol. 2. London: W. & R. Chambers.

Chaney, David. 1996. *Lifestyles*. London: Routledge.

Coleridge, Ernest Hartley, ed. 1895. *Letters of Samuel Taylor Coleridge*. Vol. I. Boston: Houghton, Mifflin and Company.

Conley, Dalton. "Cell Phones Weights Down Backpack of Self-discovery." Web: http://www.bloomberg.com/news/2011-08-30/cell-phone-burdens-backpack-of-self-discovery-commentary-by-dalton-conley.html. Accessed 20 May 2013.

Crang, Philip, Peter Jackson. 2001. "Geographies of Consumption," in *British Cultural Studies*. Eds. David Morley, Kevin Robins. Oxford: Oxford University Press.

Craze, Richard. 2004. *Out of Your Townie Mind. The Reality Behind the Dream of Country Living*. Great Ambrook: White Adder Press.

Creswell, Tim. 2006. *On the Move. Mobility in the Modern Western World*. New York: Routledge.

Davidson, Robert A. 2007. "*Terroir* and Catalonia," *Journal of Catalan Studies*.

Davis, Andrew Jackson. 2009 [1871]. "Mercurial Brainism of the Present Epoch," in *Speed Limits*. Ed. Jeffrey T. Schnapp. Milan: Skira.

De Blasi, Marlena. 2008 [2005]. *A Thousand Days in Tuscany*. London: Virago.

De Geus, Marius. 2009. "Sustainable Hedonism: The Pleasures of Living within Environmental Limits," in *The Politics and Pleasures of Consuming Differently*. Eds. Kate Soper, Martin Ryle, Lyn Thomas. Basingstoke: Palgrave Macmillan.

De Quincey, Thomas. 1862. *The Works of Thomas De Quincey: The English Mail Coach*. Vol. 4. Edinburgh: Adam and Charles Black.

Dickinson, Janet, Les Lumsdon. 2010. *Slow Travel and Slow Tourism*. London: Erthscan.

Dickinson, Janet, Les M. Lumsdon, Derek Robbins. 2011. "Slow Travel: Issues for Tourism and Climate Change." *Journal of Sustainable Tourism*, Vol. 19, No. 3.

Dillard, Annie. 1974. *Pilgrim at Tinker Creek*. New York: Harper's Magazine Press.

Duffy, Enda. 2009. *The Speed Handbook. Velocity, Pleasure, Modernism*. Durham: Duke University Press.

Elgin, Duane. 1993. *Voluntary Simplicity. Toward a Way of Life That is Outwardly Simple and Inwardly Rich*. New York: Quill.

Eriksen, Thomas Hyllard. 2001. *Tyranny of the Moment. Fast and Slow Time in the Information Age*. London: Pluto Press.

Eyres, Harry. "London's First Slow Down Festival." Web: http://www.ft.com/cms/s/0/b70c963a-1a5d-11de-9f91-0000779fd2ac.html. Accessed 10 January 2014.

Featherstone, Mike. 2005. "Introduction," in *Automobilities*. Eds. Mike Featherstone, Nigel Thrift, John Urry. London: Sage.

Ferriss, Susan, Mallory Young. 2006. *Chick Lit. The New Women's Fiction*. Eds. Susan Ferriss, Mallory Young. New York: Routledge.

Freund, Peter. 1993. *The Ecology of the Automobile*. Montreal: Black Rose Books.

Gandolfi, Franco. 2008. "The Downshifting Phenomenon," in *Downshifting. A Theoretical and Practical Approach to Living a Simple Life*. Eds. Franco Gandolfi, Hélène Cherrier. Hyderabad: The Icfai University Press.

Gardner, Nicky. 2009. "A Manifesto for Slow Travel." *Hidden Europe*. No. 25, March/April.

Gaudelli William, Timothy Patterson. 2013. "It's *Just* Geography: Critical Geography and a Critique of Advanced Placement Human Geography," in *Geography and Social Justice in the Classroom*. Ed. Todd W. Kenreich. New York: Routledge.

Gehl, Jan. 2001. *Life between Buildings*. Trans. Jo Koch. Copenhagen: Arkitektens Forlag.

Gehl, Jan, Lars Gemzøe. 2004. *Public Spaces, Public Life*. Copenhagen. Trans. Karen Steenhard. Copenhagen: Narayana Press.

Gehl, Jan, Lars Gemzøe, Sia Kirknæs, Britt Sternhagen Søndergaard. 2006. *New City Life*. Trans. Karen Steenhard. Copenhagen: The Danish Architectural Press.

George, Susan. 1998. "Fast Castes," in *Speed – Visions of an Accelerated Age*. Eds. Jeremy Millar, Michiel Schwartz. London: The Photographers' Gallery and The Trustees of the Whitechapel Art Gallery.

Giddens, Anthony. 1991. *Modernity and Self-Identity. Self and Society in the Late Modern Age*. Stanford, CA.: Stanford University Press.

Gleick, James. 1999. *Faster: The Acceleration of Just About Everything*. New York: Pantheon Books.

Goldberg, Carey. 1995. "Choosing the Joys of a Simplified Life." *The New York Times*, September 21.

Griffiths, Jay. 1999. *Pip Pip*. London: Flamingo. An Imprint of Harper Collins Publishers.

Gregg, Richard. 2009 [1936]. *The Value of Voluntary Simplicity*. Auckland: The Floating Press.

Hainsworth, Tessa. 2009. *Up with the Larks. Starting Again in Cornwall*. London: Preface Publishing.

Hassan, Robert. 2009. *Empires of Speed*. Leiden: Brill.

Henley, William Ernest. 1903. *A Song of Speed*. London: Long Acre.

Highmore, Ben. 2002. "Introduction: Questioning Everyday Life," in *The Everyday Life Reader*. London: Routledge.

Hobsbawm, Eric. 1975. *The Age of Capital 1848–75*. London: Weidenfeld and Nicolson.

Hobsbawm, Eric. 1994. *The Age of Extremes: The Short Twentieth Century, 1914–1991*. London: Michael Joseph.

Honoré, Carl. 2004. *In Praise of Slowness. Challenging the Cult of Speed*. New York: Harper One.

Honoré, Carl. 2008. "Recession? The Perfect Time to Slow Down." *The Guardian*, July 24.

Hooker, Jeremy. 1999. "Ditch Vision." *The Powys Journal* 9.

Huggins-Cooper, Lynn. 2004. *Downshift to the Good Life*. Oxford: Infinite Ideas Limited.

Illich, Ivan, Matthias Rieger, Sebastian Trapp. 1998. "Speed? What Speed?"in *Speed – Visions of an Accelerated Age*. Eds. Jeremy Millar, Michiel Schwartz. London: The Photographers'Gallery and The Trustees of the Whitechapel Art Gallery.

Illich, Ivan. 1974. *Energy and Equity*. New York: Harper & Row.

Ingold, Tim. 2000. *The Perceptions of the Environment*. London: Routledge.

Judt, Tony. 2011. *The Memory Chalet*. London: Vintage Books.

Keats, John. 2005. *Selected Letters of John Keats*. Ed. Grant F. Scott. Cambridge, Mass.: Harvard University Press.

Kern, Stephen. 1983. *The Culture of Time and Space 1880–1918*. Cambridge, MA.: Harvard University Press.

Kerridge, Richard. 2009. "Green Pleasures," in *The Politics and Pleasures of Consuming Differently*. Eds. Kate Soper, Martin Ryle, Lyn Thomas. Basingstoke: Palgrave Macmillan.

Kingwell, Mark. 1998. "Fast Forward," in *Speed – Visions of an Accelerated Age*. Eds. Jeremy Millar, Michiel Schwartz. London: The Photographers' Gallery and The Trustees of the Whitechapel Art Gallery.

Kita, Barbara. 2003. *Między przestrzeniami. O kulturze nowych mediów*. Kraków: Rabid.

Klein, Naomi. 2000. *No Logo*. London: Harper Collins.

Klein, Stefan. 2007. *The Secret Pulse of Time. Making Sense of Life's Scarcest Commodity*. Trans. Shelley Frish. Frankfurt am Main: Da Capo Press.

Kundera, Milan. 1996. *Slowness*. Trans. Linda Asher. London: Faber and Faber.

Larbaud, Valéry. 2009 [1930]. "Slowness," in *Speed Limits*. Ed. Jeffrey T. Schnapp. Trans. Christy Wampole. Milan: Skira.

Le Corbusier, 1971 [1924]. *The City of To-morrow and Its Planning*. Trans. Frederick Etchells. London: The Architectural Press.

Left, Cathy. 2009. "Preface," in *Speed Limits*. Ed. Jeffrey T. Schnapp. Milan: Skira.

Lipovetsky, Gilles. 2005. *Hypermodern Times*. Cambridge: Polity.

Lucas, Clay. "Euro-style Bike Lanes Plan for City," *The Age*. Web: http://www.theage.com.au/news/national/eurostyle-bike-lanes-planforcity/2006/09/02/1156817151269.html. Accessed 23 April 2014.

Lumsdon, Les M., Peter McGrath. 2011. "Developing a Conceptual Framework for Slow Travel: A Grounded Theory Approach." *Journal of Sustainable Tourism*, Vol. 19, No. 3.

Lyon, Janet. 1999. *Manifestoes: Provocations of the Modern*. New York: Cornell University Press.

Maeterlinck, Maurice. 2009 [1904]."In an Automobile," in *Speed Limits*. Ed. Jeffrey T. Schnapp. Trans. A. Teixeira De Mattos. Milan: Skira.

Maitland, Sara. 2009. *A Book of Silence*. London: Granta.

Malnar, Joy Monice, Frank Vodvarka. 2004. *Sensory Design*. Minneapolis: University of Minnesota Press.

Mann, Thomas. 1996 [1924]. *The Magic Mountain*. Trans. John E. Woods. New York: Vintage.

Marinetti, Filippo Tommaso. 1973. "The Founding and Manifesto of Futurism," in *Documents of 20th Century Art: Futurist Manifestoes*. Ed. Apollonio Umbro. Trans. Robert Brain, R.W Flint, J.C Higgitt, Caroline Tisdall. New York: The Viking Press.

Marinetti, Filippo Tommaso. 2002. *Selected Poems and Related Prose*. Ed. Luce Marinetti. Trans. Elizabeth R. Napier, Barbara R. Studholme. New Haven: Yale University Press.

Marinetti, Filippo Tommaso. 2009. "Le Futurism," in *Futurism. An Anthology*. Eds. Lawrence Rainey, Christine Poggi, Laura Wittman. New Haven: Yale University Press.

Marinetti, Filippo Tommaso. 2009. "The New Religion-Morality of Speed," in *Futurism. An Anthology*. Eds. Lawrence Rainey, Christine Poggi, Laura Wittman. New Haven: Yale University Press.

Maushart, Susan. 2011. *The Winter of Our Disconnect. How One Family Pulled the Plug on Their Technology and Lived to Tell/Text/Tweet the Tale*. London: Profile Books.

Merleau-Ponty, Maurice. 1962. *Phenomenology of Perception*. Trans. Colin Smith. London: Routledge & Kegan Paul.

Millar, Jeremy, Michiel Schwartz. 1998. "Introduction – Speed as a Vehicle," in *Speed – Visions of an Accelerated Age*. Eds. Jeremy Millar, Michiel Schwartz. London: The Photographers' Gallery and The Trustees of the Whitechapel Art Gallery.

Millar, Jeremy. 1998. "Rejectamenta," in *Speed – Visions of an Accelerated Age*. Eds. Jeremy Millar, Michiel Schwartz. London: The Photographers' Gallery and the Trustees of the Whitechapel Art Gallery.

Milne, Esther. "'The Minister of Locomotion': Some Historical Speculations on Velocity Culture." *M/C: A Journal of Media and Culture* 3.3 (2000).

Web: http://www.api-network.com/mc/0006/ministers.php. Accessed 13 October 2012.

Moran, Joe. 2004. "Housing, Memory and Everyday Life in Contemporary Britain." *Cultural Studies*, Vol. 18, No. 4.

Morand, Paul. 2009 [1929]. "On Speed," in *Speed Limits*. Ed. Jeffrey T. Schnapp. Trans. Jeffrey T. Schnapp. Milan: Skira.

Morasso, Mario. 2009 [1902]. "Sensations of Speed," in *Speed Limits*. Ed. Jeffrey T. Schnapp. Trans. Christy Wampole. Milan: Skira

Morse, Trent. "Slow Down, You Look Too Fast," *ARTnews*. Web: http://www.artnews.com/2011/04/01/slow-down-you-look-too-fast/. Accessed 26 April 2014.

Mowat, Robert Balmain. 1935. *Americans in England*. Boston: Houghton Mifflin Company.

Nitka, Małgorzata. 2006. *Railway Defamiliarization. The Rise of Passengerhood in the Nineteenth Century*. Katowice: Wydawnictwo Uniwersytetu Śląskiego.

Novotny, Helga. 1994 [2005]. *Time. The Modern and Postmodern Experience*. Trans. Neville Plaice. Cambridge: Polity Press.

Oldenburg, Ray. 1999. *The Great Good Place*. New York: Da Capo Press.

Oldenburg, Ray. "Our Vanishing 'Third Places,'" Planners Web. Web: http://plannersweb.com/1997/01/our-vanishing-third-places. Accessed 23 April 2014.

Orr, David W. 1992. *Ecological Literacy: Education and the Transition to a Postmodern World*. Albany: State University of New York Press.

Orr, David W. 2004. *Earth in Mind: On Education, Environment, and the Human Prospect*. Washington: Island Press.

Osbaldiston, Nick. 2013. "Conclusion: Departing Notes on the Slow Narrative,"in *Culture of the Slow. Social Deceleration in an Accelerated World*. Ed. Nick Osbaldiston. London: Palgrave Macmillan.

Osbaldiston, Nick. 2013. "Consuming Space Slowly: Reflections on Authenticity, Place and The Self," in *Culture of the Slow. Social Deceleration in an Accelerated World*. Ed. Nick Osbaldiston. London: Palgrave Macmillan.

Osbaldiston, Nick. 2013. "Slow Culture: An Introduction," in *Culture of the Slow. Social Deceleration in an Accelerated World*. Ed. Nick Osbaldiston. London: Palgrave Macmillan.

Overstreet, Harry Allen. 1969. *A Guide to Civilized Leisure*. New York: Norton & Company.

Pallasmaa, Juhani. 2005. "Lived Space: Embodied Experience and Sensory Thought," in *Encounters: Juhani Pallasmaa - Architectural Essays*. Ed. Peter MacKeith. Hämeenlinna: Rakennustieto Publishing.

Palmer, Mark. 2008. "How to Go Slow." *The Telegraph*, July 26, Web: http://www.telegraph.co.uk/news/features/3637476/How-to-go-slow.html. Accessed 28 March 2014.

Parkes, Carl, "The History of Travel Guide Books." Web: http://travelwriters.blogspot.com/2006/04/history-of-travel-guidebooks.html. Accessed 27 January 2014.

Parkins, Wendy. 2004. "At Home in Tuscany: Slow Living and the Cosmopolitan Subject." *Home Cultures*, Vol. I, Issue 3.

Parkins, Wendy. 2004. "Out of Time: Fast Subjects and Slow Living." *Time & Society*, Vol. 13, No. 2/3.

Parkins, Wendy, Geoffrey Craig. 2006. *Slow Living*. Oxford: Berg.

Petrini, Carlo. 2003. *Slow Food. The Case for Taste*. Trans. William McCuaig. New York: Columbia University Press.

Pink, Sarah. 2007. "Sensing Cittáslow: Slow Living and The Constitution of the Sensory City." *Senses & Sociology*, Vol. 2, No. 1.

Pink, Sarah. 2009. *Doing Sensory Ethnography*. London: Sage.

Pound, Ezra. 2005 [1910]. *The Spirit of Romance*. New York: New Directions.

Proust, Marcel. 2009 [1907]. "Motoring Days," in *Speed Limits*. Ed. Jeffrey T. Schnapp. Trans. Jeffrey T. Schnapp. Milan: Skira.

Ragusa, Angela T. 2013. "Downshifting or Conspicuous Consumption? A Sociological Examination of Treechange as a Manifestation of Slow Culture," in *Culture of the Slow. Social Deceleration in an Accelerated World*. Ed. Nick Osbaldiston. London: Palgrave Macmillan.

Redden, Guy, Sean Aylward Smith. 2000. "Editorial: 'Speed.'" *M/C: A Journal of Media and Culture* 3.3. Web: http://www.apinetwork.com/mc/0006/edit.php. Accessed 13 June 2011.

Relph, Edward. 1976. *Place and Placelessness*. London: Pion.

Richard Gregg. 2009. *The Value of Voluntary Simplicity*. Auckland: The Floating Press.

Ryle, Martin, Kate Soper. 2013. "Alternative Hedonism: The World by Bicycle," in *Culture of the Slow. Social Deceleration in an Accelerated World*. Ed. Nick Osbaldiston. London: Palgrave Macmillan.

Sachs, Wolfgang. 1998. "Speed Limits," in *Speed – Visions of an Accelerated Age*. Eds. Jeremy Millar, Michiel Schwartz. London: The Photographers' Gallery and The Trustees of the Whitechapel Art Gallery .

Sangster, Chris, Gillean Sangster. 2004. *The Downshifter's Guide to Relocation: Escape to a Simpler, Less Stressful Life*. Oxford: How To Books Ltd.

Sansot, Pierre. 2009 [1988]. "On the Proper Use of Slownes," in *Speed Limits*. Ed. Jeffrey T. Schnapp. Trans. Christy Wampole, Jeffrey T. Schnapp. Milan: Skira.

Sassatelli, Roberta. 2011. "Consumer Identities," in *Routledge Handbook of Identity Studies*. Ed. Anthony Elliot. New York: Routledge.

Schafer, Andreas, David G. Victor. 2000. "The Future Mobility of the World Population." *Transportation Research*, Part A 34.

Schnapp, Jeffrey T. 1999. "Crash (Speed as Engine of Individuation)." *Modernism/modernity*, Vol. 6, No. 1.

Schnapp, Jeffrey T. 2009. "Fast (Slow) Modern," in *Speed Limits*. Ed. Jeffrey T. Schnapp. Milan: Skira.

Schnapp, Jeffrey T. "Spinners." *West 86th. A Journal of Decorative Arts, Design History and Material Culture*. Web: http://www.west86th.bgc.bard.edu/articles/spinners.html. Accessed 15 June 2011.

Sennett, Richard. 1994. *Flesh and Stone. The Body and the City in Western Civilization*. New York: W.W. Norton & Company.

Severini Gino. 2009. "Plastic Analogies of Dynamism: Futurist Manifesto," in *Futurism. An Anthology*. Eds. Lawrence Rainey, Christine Poggi, Laura Wittman. New Haven: Yale University Press.

Shaw, Jenny. 2001. "'Winning territory.' Changing place to change pace," in *Timespace. Geographies of Temporality*. Eds. Jon May, Nigel Thrift. London: Routledge.

Shuffelton, George, ed. 2008. *Codex Ashmole 61: A Compilation of Popular Middle English Verse*. Kalamazoo, MI.: Medieval Institute Publications.

Smollet, Tobias. 1995 [1771]. *The Expedition of Humphry Clinker*. Ware, Hertfordshire: Wordsworth Classics.

Smyczyńska, Katrzyna. 2011. "'The Curse of the Pink Cover': Chick lit in Chick Culture," in *Beyond 2000. The Recent Novel in English*. Ed. Ewa Rychter. Wałbrzych: Wydawnictwo Państwowej Wyższej Szkoły Zawodowej.

Solnit, Rebecca. 2001. *Wanderlust. A History of Walking*. London: Verso.

Solomon, Juliet, 2009. "Happiness and the Consumption of Mobility," in *The Politics and Pleasures of Consuming Differently*. Eds. Kate Soper, Martin Ryle, Lyn Thomas. Basingstoke: Palgrave Macmillan.

Soper, Kate. 1998. "Interview. An Alternative Hedonism." Interviewed by Ted Benton. *Radical Philosophy* 92.

Soper, Kate. 2000 [1995]. *What is Nature? Culture, Politics and the Non-Human*. Oxford: Blackwell.

Soper, Kate. 2007. "Rethinking the 'Good Life': The Citizenship Dimension of Consumer Disaffection with Consumerism." *Journal of Consumer Culture*, Vol. 7, No. 2.

Soper, Kate. 2007. "The Other Pleasures of Post-Consumerism." *Soundings. A Journal of Politics and Culture*, Issue 35. Web: http://www.lwbooks.co.uk. Accessed 15 August 2013.

Soper, Kate. 2008. "Alternative Hedonism, Cultural Theory and the Role of Aesthetic Revisioning." *Cultural Studies*, Vol. 22, No. 5.

Soper, Kate. 2009. "Introduction: The Mainstreaming of Counter-Consumerist Concern," in *The Politics and Pleasures of Consuming Differently*. Ed. Kate Soper, Martin Ryle, Lyn Thomas. Basingstoke: Palgrave Macmillan.

Soper, Kate. 2010. "Humanities Can Promote Alternative 'Good Life.'" *The Guardian*, November 30. Web: http://www.theguardian.com/commentisfree/2010/nov/30/humanities-promote-alternative-good-life. Accessed 30 May 2013.

Stegemann, Michael. 1982 [Liner notes]. "A Kind of Autumnal Repose." *The Glenn Gould Edition. Goldberg Variations*. CD. Sony Classical.

St. James, Elaine. 1994. *Simplify Your Life. 100 Ways to Slow Down and Enjoy the Things that Really Matter*. New York: Hyperion.

St. James, Elaine. 1995. *Inner Simplicity. 100 Ways to Regain Peace and Nourish Your Soul*. New York: Hyperion.

St. James, Elaine. 1997. *Simplicity. Easy Ways to Simplify and Enrich Your Life*. London: Thorsons.

Stein, Jeremy. 2001. "Reflections on Time, Time-Space Compression and Technology in the Nineteenth Century," in *Timespace. Geographies of Temporality*. Eds. John May, Nigel Thrift. London: Routledge.

Stein, Karen. 2011. "Getting Away from It All: The Construction and Management of Temporary Identities on Vacation." *Symbolic Interaction*, Vol. 34, Issue 2.

Storey, John. 1994. "Introduction: The Study of Popular Culture and Cultural Studies," in *Cultural Theory and Popular Culture: A Reader*. Ed. John Storey. Harlow: Pearson Education Limited.

Synnott, Anthony. 1997. *The Body Social. Symbolism, Self and Society*. London: Routledge.

Szymborska, Wisława. 2010. *Here*. Trans. Clare Cavanagh, Stanisław Barańczak. Boston: Houghton Mifflin Harcourt.

Taylor, Joseph. 1874. *A Fast Life on the Modern Highway: Being a Glance Into the Railroad World from a New Point of View*. New York: Harper & Brothers Publishers.

Thomas, Lyn. 2008. "Alternative Realities. Downshifting Narratives in Contemporary Lifestyle Television." *Cultural Studies*, Vol. 22, No. 5.

Thomas, Lyn. 2009. "Ecochic: Green Echoes and Rural Retreats in Contemporary Lifestyle Magazines," in *The Politics and Pleasures of Consuming Differently*. Eds. Kate Soper, Martin Ryle, Lyn Thomas. Basingstoke: Palgrave Macmillan.

Tokarczuk, Olga. 2012. *Moment niedźwiedzia*. Warszawa: Wydawnictwo Krytyki Politycznej.

Tomlinson, John. 2007. *The Culture of Speed. The Coming of Immediacy*. Los Angeles: Sage.

Tsuji, Shin'ichi. 2009. "Slow is Beautiful: Culture as Slowness," in *Speed Limits*. Ed. Jeffrey T. Schnapp. Milan: Skira.

Tuan, Yi-Fu. 1977. *Space and Place: The Perspective of Experience*. Minneapolis: University of Minnesota Press.

Turner, Graeme. 2003. *British Cultural Studies. An Introduction*. London: Routledge.

Urry, John. 1990. *The Tourist Gaze. Leisure and Travel in Contemporary Societies*. London: Sage Publications.

Urry, John. 2000. *Sociology Beyond Societies. Mobilities for the Twenty-first Century*. London: Routledge.

Urry, John. 2007. *Mobilities*. Cambridge: Polity Press.

Virilio, Paul. 1999 [1997]. *Open Sky*. Trans. Julie Rose. London: Verso.

Virilio, Paul. 2006 [1977]. *Speed and Politics*. Trans. Marc Polizzotti. Los Angeles: Semiotext(e).

Virillo, Paul. 2007 [1984]. *Negative Horizon. An Essay in Dromoscopy*. London: Continuum.

Wheeler, Karen. 2008. *Tout Sweet: Hanging Up my High Heels for a New Life in France*. Chichester: Summersdale.

Williams, Raymond. 1993 [1973]. *The Country and the City*. London: The Hogarth Press.

Willis, Paul. 1979. "Shop Floor Culture, Masculinity and the Wage Form," in *Working Class Culture: Studies in History and Theory*. Eds. John Clarke, Chas Critcher, Richard Johnson. London: Hutchinson.

Zardini, Mirko. 2009. "Preface," in *Speed Limits*. Ed. Jeffrey T. Schnapp. Milan: Skira.

Zimbardo, Philip, John Boyd. 2008. *The Time Paradox. The New Psychology of Time*. London: Rider.

Web sites

http://www.affirmpress.com.au

http://www.artnews.com

http://www.cittaslow.org

http:// www.cittaslow.net

http://www.craftspace.co.uk

http://www.ft.com

http://www.guardian.co.uk

http://www.lwbooks.co.uk

http://www.slowfood.com

http://www.slowhomestudio.com

http://www.slowmovement.com

http://www.slowplanet.com

http://www.slowtrav.com

http://www.theage.com.au

http://www.telegraph.co.uk

http://www.travelwriters.blogspot.com

http://www.worldwidewords.org

Literary and Cultural Theory

General editor: Wojciech H. Kalaga

Vol. 1 Wojciech H. Kalaga: Nebulae of Discourse. Interpretation, Textuality, and the Subject. 1997.

Vol. 2 Wojciech H. Kalaga / Tadeusz Rachwał (eds.): Memory – Remembering – Forgetting. 1999.

Vol. 3 Piotr Fast: Ideology, Aesthetics, Literary History. Socialist Realism and its Others. 1999.

Vol. 4 Ewa Rewers: Language and Space: The Poststructuralist Turn in the Philosophy of Culture. 1999.

Vol. 5 Floyd Merrell: Tasking Textuality. 2000.

Vol. 6 Tadeusz Rachwał / Tadeusz Slawek (eds.): Organs, Organisms, Organisations. Organic Form in 19th-Century Discourse. 2000.

Vol. 7 Wojciech H. Kalaga / Tadeusz Rachwał: Signs of Culture: Simulacra and the Real. 2000.

Vol. 8 Tadeusz Rachwal: Labours of the Mind. Labour in the Culture of Production. 2001.

Vol. 9 Rita Wilson / Carlotta von Maltzan (eds.): Spaces and Crossings. Essays on Literature and Culture in Africa and Beyond. 2001.

Vol. 10 Leszek Drong: Masks and Icons. Subjectivity in Post-Nietzschean Autobiography. 2001.

Vol. 11 Wojciech H. Kalaga / Tadeusz Rachwał (eds.): Exile. Displacements and Misplacements. 2001.

Vol. 12 Marta Zajac: The Feminine of Difference. Gilles Deleuze, Hélène Cixous and Contemporary Critique of the Marquis de Sade. 2002.

Vol. 13 Zbigniew Bialas / Krzysztof Kowalczyk-Twarowski (eds.): Alchemization of the Mind. Literature and Dissociation. 2003.

Vol. 14 Tadeusz Slawek: Revelations of Gloucester. Charles Olsen, Fitz Hugh Lane, and Writing of the Place. 2003.

Vol. 15 Carlotta von Maltzan (ed.): Africa and Europe: En/Countering Myths. Essays on Literature and Cultural Politics. 2003.

Vol. 16 Marzena Kubisz: Strategies of Resistance. Body, Identity and Representation in Western Culture. 2003.

Vol. 17 Ewa Rychter: (Un)Saying the Other. Allegory and Irony in Emmanuel Levinas's Ethical Language. 2004.

Vol. 18 Ewa Borkowska: At the Threshold of Mystery: Poetic Encounters with Other(ness). 2005.

Vol. 19 Wojciech H. Kalaga / Tadeusz Rachwał (eds.): Feeding Culture: The Pleasures and Perils of Appetite. 2005.

Vol. 20 Wojciech H. Kalaga / Tadeusz Rachwał (eds.): Spoiling the Cannibals' Fun? Cannibalism and Cannibalisation in Culture and Elsewhere. 2005.

Vol. 21 Katarzyna Ancuta: Where Angels Fear to Hover. Between the Gothic Disease and the *Meat*aphysics of Horror. 2005.

Vol. 22 Piotr Wilczek: (Mis)translation and (Mis)interpretation: Polish Literature in the Context of Cross-Cultural Communication. 2005.

Vol. 23 Krzysztof Kowalczyk-Twarowski: *Glebae Adscripti*. Troping Place, Region and Nature in America. 2005.

Vol.	24	Zbigniew Białas: The Body Wall. Somatics of Travelling and Discursive Practices. 2006.
Vol.	25	Katarzyna Nowak: Melancholic Travelers. Autonomy, Hybridity and the Maternal. 2007.
Vol.	26	Leszek Drong: Disciplining the New Pragmatism. Theory, Rhetoric, and the Ends of Literary Study. 2007.
Vol.	27	Katarzyna Smyczyńska: The World According to Bridget Jones. Discourses of Identity in Chicklit Fictions. 2007.
Vol.	28	Wojciech H. Kalaga / Marzena Kubisz (eds.): Multicultural Dilemmas. Identity, Difference, Otherness. 2008.
Vol.	29	Maria Plochocki: Body, Letter, and Voice. Construction Knowledge in Detective Fiction. 2010.
Vol.	30	Rossitsa Terzieva-Artemis: Stories of the Unconscious: Sub-Versions in Freud, Lacan and Kristeva. 2009.
Vol.	31	Sonia Front: Transgressing Boundaries in Jeanette Winterson's Fiction. 2009.
Vol.	32	Wojciech Kalaga / Jacek Mydla / Katarzyna Ancuta (eds.): Political Correctness. Mouth Wide Shut? 2009.
Vol.	33	Paweł Marcinkiewicz: The Rhetoric of the City: Robinson Jeffers and A. R. Ammons. 2009.
Vol.	34	Wojciech Małecki: Embodying Pragmatism. Richard Shusterman's Philosophy and Literary Theory. 2010.
Vol.	35	Wojciech Kalaga / Marzena Kubisz (eds.): Cartographies of Culture. Memory, Space, Representation. 2010.
Vol.	36	Bożena Shallcross / Ryszard Nycz (eds.): The Effect of Pamplisest. Culture, Literature, History. 2011.
Vol.	37	Wojciech Kalaga / Marzena Kubisz / Jacek Mydla (eds.): A Culture of Recycling / Recycling Culture? 2011.
Vol.	38	Anna Chromik: Disruptive Fluidity. The Poetics of the Pop *Cogito*. 2012.
Vol.	39	Paweł Wojtas: Translating Gombrowicz's Liminal Aesthetics. 2014.
Vol.	40	Marcin Mazurek: A Sense of Apocalypse. Technology, Textuality, Identity. 2014.
Vol.	41	Charles Russell / Arne Melberg / Jarosław Płuciennik / Michał Wróblewski (eds.): Critical Theory and Critical Genres. Contemporary Perspectives from Poland. 2014.
Vol.	42	Marzena Kubisz: Resistance in the Deceleration Lane. Velocentrism, Slow Culture and Everyday Practice. 2014

www.peterlang.com

www.ingramcontent.com/pod-product-compliance
Ingram Content Group UK Ltd.
Pitfield, Milton Keynes, MK11 3LW, UK
UKHW041923210426
5322IPUK00002B/22